NIST Technical Note 1681

Best Practice Guidelines for Structural Fire Resistance Design of Concrete and Steel Buildings

Long T. Phan
Therese P. McAllister
John L. Gross
Engineering Laboratory
National Institute of Standards and Technology

Morgan J. Hurley
Society of Fire Protection Engineers

November 2010

U.S. Department of Commerce
Gary Locke, Secretary

National Institute of Standards and Technology
Patrick D. Gallagher, Director

Abstract

This document is intended to provide practicing engineers and building code officials with a technical resource that contains the current "best practice" for fire-resistant design of concrete and steel structures. The report provides a review of existing U.S. and international guidelines and design standards, which use approaches that range from simple prescriptive methods to sophisticated software programs with advanced methods of analysis under a wide range of realistic fire conditions. Basic concepts of risk-informed decision making for mitigating fire risk, and a general framework for assessing fire risk to building construction and for developing structural design requirements for fire conditions are described. Current best knowledge in thermal and mechanical properties and behaviors of normal strength concrete, high strength concrete, structural steel, and several major groups of common fire protection materials at elevated temperatures, which are necessary for performance-based engineering calculation, are presented. Modern fire-resistant design methodologies for concrete and steel structures are discussed, including methods based on standard fire tests as well as performance-based engineering analysis methods that involve heat transfer and structural analysis at elevated temperatures.

This report is not intended to provide step-by-step design procedures. Rather, it provides general guidance on the approaches to, and practical aspects of, implementing a fire-resistant design approach for concrete and steel buildings. The guidance includes key concepts and examples for identifying performance objectives, conducting risk analyses, selecting design fire scenarios and fire exposure curves, and implementing heat transfer and structural response analyses for the structural fire-resistant design of concrete and steel structures.

The best practice guidelines document is intended to be a technical resource that provides the most complete information for structural fire-resistant design at present. The document is a non-consensus, non-mandatory document and, as such, it does not represent a minimum standard of care.

Keywords: building codes; concrete structures; design fire scenarios; fire-resistant design; fire risk mitigation; performance-based fire engineering; steel structures; structural response analysis; thermal analysis.

Disclaimers:

(1) The policy of the NIST is to use the International System of Units in its technical communications. In this document however, works of authors outside NIST are cited which describe measurements in certain non-SI units. Thus, it is more practical to include the non-SI unit measurements from these references.

(2) Certain trade names or company products are mentioned in the text to specify adequately the experimental procedure and equipment used. In no case does such identification imply recommendation or endorsement by the National Institute of Standards and Technology, nor does it imply that the products are the best available for the purpose.

Preface

This best practices document is prepared as part of the response to the recommendations from the October 2003 NIST-SFPE Workshop for Development of a National R&D Roadmap for Structural Fire Safety Design and Retrofit of Structures. The workshop was jointly organized by the Society of Fire Protections Engineers (SFPE) and the National Institute of Standards and Technology (NIST) with guidance from an industry steering team comprising of representatives of the American Institute of Steel Construction (AISC), the American Society of Civil Engineers (ASCE), Portland Cement Association (PCA), Underwriters Laboratories (UL), Arup Fire, and National Fire Protection Association (NFPA). The document aims to integrate information on current "best practices" in fire resistance engineering in the United States and overseas with current best knowledge in fire risk assessment and characterization, material properties and responses at high temperatures, and thermal and structural response calculation methods into a single source document, and to provide engineers and code enforcement officials with a technical resource for use in their approach to fire resistance design and evaluation of concrete and steel structures.

This best practices guidance document provides the current best information to enable the use of the performance-based design approach for fire resistance design of those structures whose performance objectives might exceed the objectives intended by current codes. It is a non-consensus, non-mandatory document. As such, it does not represent a minimum standard of care and is not intended to supersede any existing standards currently in practice. Where current best practices are represented in existing standards, the standards are discussed and referenced.

The preparation of this document was coordinated by Mr. Bernard Murphy of the Multihazard Mitigation Council (MMC) of the National Institute of Building Sciences (NIBS). Under NIST guidance, the MMC convened an industry steering group that included Daniel Falconer (American Concrete Institute, ACI), James Rossberg (ASCE), Jason Krohn (Precast/Prestressed Concrete Institute, PCI), John Ruddy (AISC, formerly Structural Affiliates International, Inc.), and Morgan Hurley (SFPE) to provide guidance in defining the scope of the document, developing the working outline, and recommending appropriate chapter authors. To that end, the MMC selected the following individuals who are experts in various fields of structural and fire safety engineering to serve as chapter authors: Farid Alfawakhiri (AISI), Andy Buchanan (University of Canterbury), Donald Dusenberry (Simpson Gumpertz & Heger Inc.), Bruce Ellingwood (Georgia Institute of Technology), Morgan Hurley (SFPE), Jim Milke (University of Maryland), Stephen Pessiki (Lehigh University), and Long Phan (NIST). Subsequently, the MMC convened an introductory planning meeting with the authors and the industry steering group, organized an internal workshop at NIBS for presentation and review of the 60 % draft document and an invited workshop in Washington, D.C. for the presentation and comments on the 90 % draft document. A select group of 65 individuals representing the engineering and construction industry was invited to attend the 90 % draft workshop and to provide comments. Following this workshop, the MMC coordinated the incorporation of comments into an interim draft document and posted the document on the NIST FTP website for further comments. Comments provided were subsequently included in this the final version of the Best Practice Guidelines for Structural Fire Resistance Design of Concrete and Steel Buildings.

Table of Contents

List of Figures

List of Tables

Chapter 1

Introduction

Long Phan, Ph.D., P.E., National Institute of Standards and Technology

1.1 BASIC EVALUATION CONCEPTS AND GOALS

Current fire protection strategy for a building often incorporates a combination of active and passive fire protection measures. Active measures, such as fire alarm and detection systems or sprinklers, require either human intervention or automatic activation and help control fire spread and its effect as needed at the time of the fire. Passive fire protection measures are built into the structural system by:

- Choice of building materials
- Dimensions of building components
- Compartmentation, and
- Fire protection materials

These control fire spread and its effect by providing sufficient *fire resistance* to prevent loss of structural stability within a prescribed time period, which is based on the building's occupancy and fire safety objectives. Materials and construction assemblies that provide fire resistance, measured in terms of fire endurance time, are commonly referred to as *fire resistance-rated-construction* or *fire-resistive materials and construction* in the current two model building codes: International Building Code 2006 (ICC 2006) and the Building Construction and Safety Code, NFPA 5000 (NFPA 2006).

Current U.S. practice for fire resistance design of structures is principally based on the provisions of locally adopted codes that are usually based on the model building codes. The codes specify minimum required fire endurance times (or fire endurance ratings) for building elements and accepted methods for determining their fire endurance ratings. The permitted methods for determining fire endurance ratings include:

1. Qualification testing, which requires the building element to be tested under a standard fire exposure (more details in Section 2.5.2) in accordance with the procedure and criteria set forth in ASTM E 119 (2007) or NFPA 251 (2006), or
2. Analytical methods as prescribed by ACI/TMS 216.1 (ACI, 2007), ASCE/SFPE 29 (2005), and ANSI/AISC 360 (2005).

ACI/TMS 216.1 (TMS: The Masonry Society) prescribes calculation methods for determining equivalent fire resistance ratings for structural members and barrier assemblies made of concrete, concrete masonry, and clay masonry that would have been equivalently achieved in the ASTM E 119 standard fire test. ASCE/SFPE 29 prescribes similar calculation methods but with additional calculation methods for steel and wood construction.

The prescriptive approach to fire safety engineering is simple to implement and enforce and is satisfactory in meeting the codes' stated intent, which is:

> "...to establish the minimum requirements to safeguard the public health, safety and general welfare through structural strength, means of egress facilities, stability, sanitation, adequate light and ventilation, energy conservation, and safety to life and property from fire and other hazards attributed to the built environment." (ICC 2006)

However, as will be further articulated in the introduction to Chapter 3 of these guidelines, there is also a growing recognition that the current prescriptive, component-based method only provides a relative comparison of how similar building elements performed under a standard fire exposure and does *not* provide information about the actual performance (i.e., load-carrying capacity) of a component or assembly in a real fire environment, nor of the system as a whole or its connections. The prescriptive method also does not provide how the structural system as a whole or its connections will perform in a standard fire exposure, nor does it account for the effects of thermal expansion on the strength and stability of a structural system. Therefore, this method cannot be used to quantify the maximum possible fire endurance time of a structure without undergoing collapse. Thus, for a certain class of buildings such as high-rises or other important structures which, due to the longer evacuation time or the significance of the buildings, may be required to survive beyond the unquantified system fire endurance time without structural collapse using prescriptive methods, a performance-based fire resistance approach may provide a more rational method for achieving the necessary fire resistance more consistent with the needed level of protection. A performance-based fire resistance approach considering the evolution of the building's structural capacity as it undergoes realistic (non-standard) fire exposures is thus a desirable alternative fire resistance design method for those structures. This approach may also be used with a standard fire if that is deemed appropriate.

Fire can affect a building's structural capacity in two ways:

1. Prolonged exposure of structural components or subsystems to elevated temperatures degrades their engineering properties, thus resulting in the reduction of overall structural capacity.
2. Exposure to elevated temperature may induce internal forces (due to restraint of thermal expansion) or axial deformations in structural members due to plastic and creep strains or buckling, which may adversely affect the global stability of the building.

For steel structures, the effect of geometric nonlinearity is likely to be significant because of large deformation that may occur. For concrete structures, lateral displacement of columns at slab–column joints due to thermal expansion of the slabs might pose additional risk to the global stability of the structure. Further, because steel has relatively small thermal mass and high thermal conductivity, temperature is more likely to be nearly uniform across most steel sections while concrete components can have steep thermal gradients near their surface, which may cause surface spalling. Consideration of the evolution of the building's structural capacity and global stability requires a performance-based fire engineering approach that explicitly considers structural fire loads in the design process to achieve a rational fire safety design.

1.2 PURPOSE OF THESE GUIDELINES

These best practice guidelines are a result of a collaborative effort initiated by NIST as part of its World Trade Center (WTC) Response Plan and through its Safety of Threatened Building R&D program. NIST collaborated with representatives of the American Concrete Institute (Daniel Falconer), the American Society for Civil Engineers (James Rossberg), the Precast/Prestressed Concrete Institute (Jason Krohn), the American Institute of Steel Construction (John Ruddy, formerly with Structural Affiliates International, Inc.), and the Society of Fire Protection Engineers (Morgan Hurley), who served as steering committee members and provided technical oversight. Individual experts in the fields of structural and fire safety engineering authored the chapters in this document:

Chapter 1	Long Phan, NIST
Chapter 2	Andy Buchanan, University of Canterbury
Chapter 3	Bruce Ellingwood, Georgia Institute of Technology
Chapter 4	Jim Milke, University of Maryland
	Stephen Pessiki, Lehigh University
	Long Phan, NIST
Chapter 5	Farid Alfawakhiri, American Iron and Steel Institute
Chapter 6	Donald Dusenberry, Simpson Gumpertz & Heger Inc.
	Morgan Hurley, Society of Fire Protection Engineers

The guidelines were produced under the coordination of Mr. Bernard Murphy of the Multi-Hazard Mitigation Council (MMC) of the National Institute of Building Sciences (NIBS). NIST, together with Morgan Hurley (SFPE), assumed the role of technical editors to provide consistency between chapters and to add additional technical information reflecting new knowledge gained in the course of the WTC investigation. Individuals who contributed in this regard include Dilip Banerjee (NIST, Building and Fire Research Laboratory) and William Luecke (NIST, Materials Science and Engineering Laboratory).

These best practice guidelines are a non-consensus, non-mandatory document. It is intended to be a technical resource that provides the most complete information for structural fire-resistant design at present. As a non-mandatory document, it does not represent a minimum standard of care.

These guidelines aim to integrate information on current practice in fire resistance engineering in the United States and overseas with current best knowledge in fire risk assessment and characterization, material responses at high temperatures, and thermal and structural response calculation methods into a single source document, and to provide engineers and code enforcement officials with a technical resource for use in their approach to fire resistance design and evaluation of concrete and steel structures. They are not intended to supersede any existing standards currently in practice. Rather, they provide the current best information to enable the use of an alternate performance-based design process for fire resistance design of those structures. Where current best practices are represented in existing standards, the standards are discussed and referenced.

1.3 SCOPE

The information presented in this document is limited to passive fire protection measures only. The material properties and calculation procedures in Chapters 4 and 5 are based on the assumption that intervention by active fire protection measures such as sprinklers or firefighters is nonexistent. While active fire protection systems are an important component of a fire protection strategy, active systems may fail and human intervention may not be timely, leaving passive fire resistance measures as the only fire protection. Since these guidelines focus only on the structural fire resistance calculation and not the larger structural fire safety strategies, the issue of "trade-offs" between active and passive fire protection measures is not considered.

The information in these guidelines is also limited to concrete and steel. Other common building construction materials such as masonry and wood are not included since they are less likely to be used as the main structural material in high-rise construction, which is the type of construction more likely to utilize the performance-based fire resistance design approach.

While some parts of these guidelines may be broadly applied (e.g., Chapter 3, Decision Framework for Fire Risk Mitigation), the guidelines may not be applicable in all instances such as for industrial buildings or chemical process plants.

Chapter 2 presents a review of existing U.S. and international guidelines and design standards for fire safety design of concrete and steel structures. It also discusses general considerations for structural design for fire conditions and the different levels of sophistication in the fire design process currently permitted by the Eurocodes.

Chapter 3 outlines the decision framework for fire risk mitigation that includes a discussion of performance objectives for fire-resistant design, fire hazard modeling, and steps involved in structural engineering for fire conditions.

Chapter 4 provides information on material properties of concrete and steel reinforcement as functions of temperature and design procedures with discussions of methods for thermal and structural analyses of cast-in-place and precast concrete construction.

Chapter 5 provides information on material properties of structural steel and fire protection materials and discusses methods for structural fire design of steel members.

Chapter 6 provides general guidance on structural fire resistant design approaches, including technical considerations for fire modeling issues, and discusses practical aspects of implementing these design approaches for applications to design of concrete and steel structures.

1.4 REFERENCES

ACI (2007), *"Code Requirements for Determining Fire Resistance of Concrete and Masonry Construction Assemblies,"* ACI/TMS 216.1-07, Farmington Hills, Michigan: American Concrete Institute.

ANSI/AISC (2005), *"Specification for Structural Steel Buildings,"* ANSI/AISC 360-05, American Institute of Steel Construction, Inc., Chicago, Illinois.

ASCE/SFPE (2005), *Standard Calculation Methods for Structural Fire Protection*, ASCE/SEI/SFPE 29-05, Reston, VA: American Society of Civil Engineers.

ASTM (2007), *Standard Test Methods for Fire Tests of Building Construction and Materials*, E119-07a, West Conshohocken, Pa.: American Society for Testing and Materials.

ICC (2006), *International Building Code*, Falls Church, Va.: International Code Council.

NFPA (2006), *Building Construction and Safety Code, NFPA 5000*, Quincy, Mass.: National Fire Protection Association.

NFPA (2006), *"Standard Methods of Tests of Fire Endurance of Building Construction and Materials,"* NFPA 251, Quincy, Mass.: National Fire Protection Association.

Chapter 2

Existing Guidelines for Fire Resistance Design of Concrete and Steel Structures

Andy Buchanan, Ph.D., University of Canterbury

2.1 INTRODUCTION

This chapter describes existing international guidelines and design standards for structural fire resistance of concrete and steel structures. These guidelines and standards range from very simple prescriptive documents to textbooks and sophisticated codes that allow advanced methods of analysis under a wide range of realistic conditions. There is focus on current U.S. developments and on the Structural Eurocodes, which are the most comprehensive suite of documents for structural fire design at the present time.

This chapter also benchmarks the international status of development and adoption of design standards for structural fire resistance of concrete and steel structures, within the overall codes and standards context.

2.2 LEGISLATIVE ENVIRONMENT

The legislative environment differs greatly among countries. Many countries are moving at various speeds to adopt performance-based codes, or to move from a prescriptive code environment to a more performance-based environment.

2.2.1 Prescriptive Codes

In the United States there are two main building codes: the International Building Code (ICC 2006) and the NFPA Building Construction and Safety Code (NFPA 2006). Both of these codes specify fire resistance ratings in a prescriptive environment. Designers have the freedom to select materials and assemblies to meet these requirements using the methods described below in Section 2.6.

2.2.2 Performance-Based Codes

Until recently, structural design for fire resistance in all countries has been based on *prescriptive* building codes, with little or no opportunity for designers to take a rational engineering approach to the provision of fire safety. Recently, many countries have adopted *performance-based* building codes, which allow designers to use any fire safety strategy they wish, provided that adequate safety can be demonstrated. In general terms, a prescriptive code states "how a building is to be constructed" whereas a performance-based code states "how a building is to perform" (Buchanan 2001). An important part of performance-based design is identification of the severity and probability of the design hazards.

In the development of new codes, many countries have adopted a multi-level approach to fire resistance design as shown in Figure 2.1. At the highest levels, there is legislation specifying the overall goals, functional objectives, and required performance that must be achieved in all buildings. At a lower level, there is a selection of alternative means of achieving those goals. The three most common options are to:

1. Comply with a prescriptive "acceptable solution"
2. Comply using an "approved calculation method"
3. Carry out a performance-based "alternative design" from engineering principles using all the information available

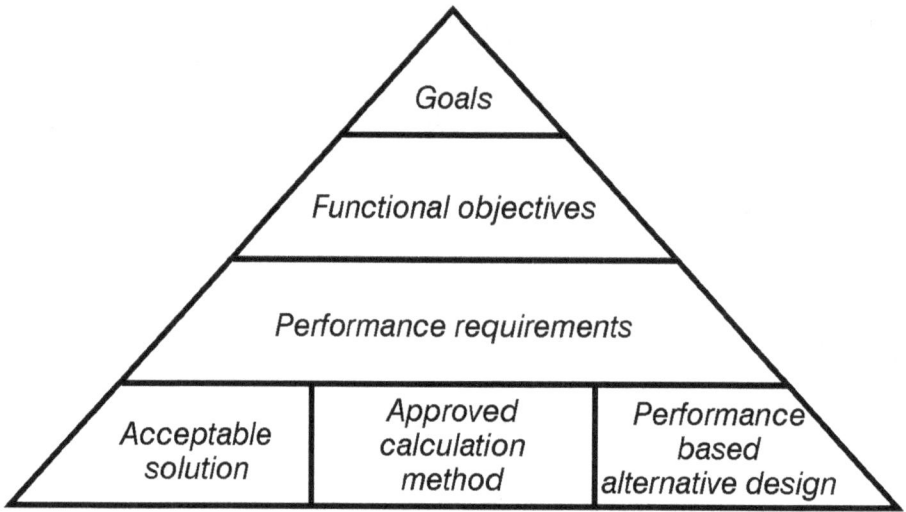

FIGURE 2.1. Hierarchical Relationship for Performance-Based Design (Buchanan 2001)

Standard calculation methods for all aspects of fire resistance design have not yet been developed for widespread use, so compliance with performance-based codes in most countries is usually achieved by simply meeting the requirements of "acceptable solutions" (a "deemed-to-satisfy" solution), or alternatively carrying out a performance-based "alternative design" based on fire engineering principles. Alternative designs can often be used to justify variations from the "acceptable solution" in order to provide cost savings or other benefits.

Codes differ around the world. They all have the objectives of protecting life and property from the effects of fire, but the emphasis between life safety and property protection varies considerably. The code environment in England, Australia, and some Scandinavian countries is similar to that in New Zealand (described by Buchanan 1994, 2000).

Moves toward performance-based codes are being taken slowly. More background on performance-based design in the U.S. context is given by Custer and Meacham (1997) and SFPE (2000). NFPA 5000 (2006) and ICC Performance Code (2006) include specific provisions that address performance-based design.

2.2.3 Eurocodes

For more than 25 years, European countries have been working on a new coordinated set of structural design standards known as the Structural Eurocodes. These are comprehensive documents that bring together diverse European views on all aspects of structural design for all main structural materials. The Eurocodes are being prepared by the European Committee for Standardization (CEN) under an agreement with Commission of the European Community. The Eurocodes recognize the need for member countries to set national safety standards that may vary from country to country, so each country's national standard will comprise the full text of the Eurocode with local modifications in a supporting document. More details are given later in this chapter.

2.3 STRUCTURAL DESIGN FOR FIRE CONDITIONS

Structural design for fire conditions is multi-faceted. It can be done in a very simple approximate way or with more and more detail considering more and more of the important variables. Whatever the level of complexity, it is essential for the designer to know what is being achieved and what assumptions are being made. A recent book giving an overview of the whole field, including fire severity, fire resistance, and design of concrete, steel, and timber structures, is given by Buchanan (2001). Some of the material in the book is summarized below in Section 2.8.3.

2.3.1 Design Objectives

For any design outside a simple prescriptive requirement, fire safety objectives must be established first. The overall design framework needs to set objectives for property protection and safety of occupants and fire fighters for a range of design hazards. Design for fire safety is often split into *active* and *passive* fire protection. A major component of passive fire protection is *fire resistance*, which is only one component of the overall fire safety strategy. Structural design for fire safety is a subset of fire resistance.

Structural elements can be provided with fire resistance for either *controlling the spread of fire* or *preventing structural collapse*, or both, depending on the functional requirements for the particular building. This chapter concentrates on the latter.

2.3.2 Design Process

Most codes recognize that structural design for fire conditions is conceptually similar to structural design for normal temperature conditions. Before making any design, it is essential to establish clear objectives and determine the severity of the design fire. The design can be carried out using either working stress or ultimate strength (LRFD) format. The main differences of fire design compared with normal temperature design are that, at the time of a fire:

- The applied loads are less.
- Internal forces may be induced by thermal expansion.
- Strengths of materials may be reduced by elevated temperatures.
- Cross section areas may be reduced by charring or spalling.
- Smaller safety factors can be used because of the low likelihood of the event.
- Deflections may be important as they may affect strength and global stability.
- Different failure mechanisms need to be considered.

The above factors may be different for different materials.

2.3.3 Loads for Structural Fire Design

The most likely loads at the time of a fire are much lower than the maximum design loads specified for normal temperature conditions. This is especially true for members that have been designed for load combinations including wind, snow, or earthquake, or for members sized for deflection control or architectural reasons. For this reason, different design loads and load combinations are used. It is generally assumed that there is no explosion or other structural damage associated with the fire. Loads on members could be much higher if other members are removed or distressed.

Most codes refer to an "arbitrary point-in-time load" to be used for the fire design condition. As an example, the ASCE document (ASCE 2005) gives the design load combination for fire U_f as

$$U_f = 1.2D_n + 0.5 L_n$$

where D_n and L_n are the design levels of dead and live load respectively, from the standard.

The Eurocode recommendations differ slightly as shown in Table 2.1, with two values: the first is for storage occupancies with semi-permanent loads such as library books or other stored items, and the second is for all other occupancies.

TABLE 2.1. Dead and Live Load Factors for Fire Design

	Dead load	Storage load	Other live load
U.S.A. (ASCE 2005)	$1.2D_n$	$0.5 L_n$	$0.5 L_n$
Ellingwood and Corotis (1991)	D_n	$0.5 L_n$	$0.5 L_n$
Eurocode (EC1 1994)	D_n	$0.9 L_n$	$0.5 L_n$
New Zealand (SNZ 1992)	D_n	$0.6 L_n$	$0.4 L_n$

2.3.4 Design Equation

The fundamental step in designing structures for fire safety is to verify that the fire resistance of the structure (or each part of the structure) is greater than the severity of the fire to which the structure is exposed. This verification requires that:

Fire resistance ≥ Fire severity

Where:
> *Fire resistance* = A measure of the ability of the structure to resist collapse, fire spread, or other failure during exposure to a fire of specified severity
>
> *Fire severity* = A measure of the destructive impact of a fire, or a measure of the forces or temperatures that could cause collapse or other failure as a result of the fire

As shown in Table 2.2, there are three alternative methods of comparing fire severity with fire resistance. The verification may be in the *time* domain, the *temperature* domain, or the *strength* domain, using different units, which can be confusing if not understood clearly. The first two domains are based on the fire resistance rating (FRR), which is the time to failure under standard fire conditions, expressed in different units but giving the same result. The third domain (*strength*) is most often used with realistic fires where it has to be shown that the structure will not fail at any point during the full process of fire development and decay.

TABLE 2.2. Three Alternative Methods of Comparing Fire Severity with Fire Resistance

Domain	Units	FIRE RESISTANCE	≥	FIRE SEVERITY
Time	minutes or hours	Time to failure (FRR)	≥	Fire duration as calculated or specified by code
Temperature	°C	Steel temperature to cause failure	≥	Maximum steel temperature reached during the fire
Strength	kN or kN.m	Load capacity (strength/stability) at elevated temperature	≥	Applied load during the fire

2.4 LEVELS OF SOPHISTICATION IN THE DESIGN PROCESS

One of the difficulties in describing or specifying fire-resistant design of structures is that there are many different levels of sophistication in the design process with regard to the part (or parts) of the structure being considered, the fire exposure, and the structural analysis under fire conditions.

Three levels of specifying design for fire performance (as described in the structural Eurocodes) are:

1. Tabulated data
2. Simplified calculation methods
3. Advanced calculation models

Three possible levels of analysis of structural behavior are:

1. Single member analysis
2. Analysis of sub-assemblage or part of the structure
3. Global structural analysis of the whole structure

Four possible levels of fire exposure are:

1. Code-specified time of exposure to the standard fire
2. Time of exposure to the standard fire equivalent to a complete burnout
3. Parametric fire based on the standard fire with a decay phase
4. Independently calculated exposure of expected real fire

Table 2.3 (adapted from EC2 2002) illustrates the applicability of three levels of specifying design for fire performance, showing where each can be used for single members, parts of a structure, or global structural analysis.

TABLE 2.3. Alternative Methods of Verifying Fire Performance

	Tabulated data	Simplified calculation methods	Advanced calculation models
Member analysis	YES Standard fire only	YES Standard fire and parametric fire	YES Parametric fire or real fire Design from first principles
Analysis of parts of the structure	NO	YES Standard fire and parametric fire	YES Parametric fire or real fire Design from first principles
Global structural analysis	NO	NO	YES Parametric fire or real fire Design from first principles

2.5 FIRE SEVERITY

2.5.1 Fire Severity for Design

The fire severity to be used for design depends on the legislative environment and on the design philosophy. In a *prescriptive* code, the design fire severity is usually prescribed by the code with little or no room for discussion. In a *performance-based* code, the design fire is usually recommended to be a complete burnout, or in some cases a shorter time of fire exposure that only allows for escape, rescue, or firefighting (Buchanan 2001). The *equivalent time* of a complete burnout is the time of exposure to the standard test fire that would result in an equivalent impact on the structure.

2.5.2 Standard Fire Exposure

Most countries rely on large-scale fire resistance tests to assess the fire performance of building materials and structural elements. The time–temperature curve used in fire resistance tests is called the *standard fire*. Full-size tests are preferred over small-scale

12

tests because they allow the method of construction to be assessed, including the effects of thermal expansion, shrinkage, local damage, and deformation under load.

The most widely used standard test specifications are ASTM E119 (ASTM 2007), NFPA 251 (NFPA 2006), UL 263 (UL 2003), and ISO 834 (ISO 1975). Other national standards include British Standard BS 476, Parts 20-23 (BSI 1987), Canadian Standard CAN/ULC-S101-04 (ULC 2004), and Australian Standard AS 1530, Part 4 (SAA 1990). The standard time–temperature curves from ASTM E119 and ISO 834 are compared in Figure 2.2. They are seen to be rather similar. All other international fire resistance test standards specify similar time–temperature curves.

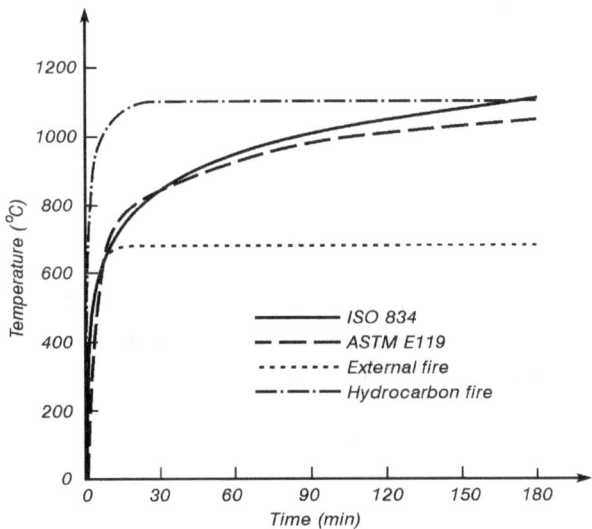

FIGURE 2.2. Standard Time–Temperature Curves

The ASTM E119 curve is defined by a number of discrete points, which are shown in Table 2.4 along with the corresponding ISO 834 temperatures. Several equations approximating the ASTM E119 curve are given by Lie (2002), the simplest of which gives the temperature T (°C) as

$$T = 750 \left[1 - e^{-3.79553\sqrt{t_h}} \right] + 170.41\sqrt{t_h} + T_0$$

Where:

t_h = Time (hours)

13

TABLE 2.4. ASTM E119 and ISO 834 Time–Temperature Curves

Time (min)	ASTM E119 Temperature (°C)	ISO 834 Temperature (°C)
0	20	20
5	538	576
10	704	678
30	843	842
60	927	945
120	1010	1049
240	1093	1153
480	1260	1257

The ISO 834 specification (ISO 1975) defines the temperature T (°C) by the following equation:

$$T = 345 \log_{10} (8t + 1) + T_0$$

Where:

t = Time (minutes)

T_0 = Ambient temperature (°C)

Figure 2.2 also shows two alternative design fires from the Eurocode (EC1 2002). The upper curve is the hydrocarbon fire curve, intended for use where a structural member is engulfed in flames from a large pool fire. The temperature T (°C) is given by

$$T = 1080 (1 - 0.325e^{-0.167t} - 0.675e^{-2.5t}) + T_0$$

Where:

t = Time (minutes)

T_0 = Ambient temperature (°C)

The lower curve is intended for design of structural members located outside a burning compartment. Unless they are engulfed in flames, exterior structural members will be exposed to lower temperatures than members inside a compartment. The temperature for external members is given by

$$T = 660 (1 - 0.687e^{-0.32t} - 0.313e^{-3.8t}) + T_0$$

2.5.3 Realistic Fire Exposure

If a fire in a typical room is allowed to grow without intervention, assuming sufficient fuel and ventilation, temperatures will increase as the radiant heat flux to all objects in the room increases. At a critical level of heat flux, all exposed combustible items in the room will ignite, leading to a rapid increase in both heat release rate and temperature. This transition is known as *flashover*, after which the fire is often referred to as a "post-flashover fire," "fully developed fire," or "full room involvement."

14

As an example of realistic fire exposure in a compartment, Figure 2.3 shows typical time–temperature curves for post-flashover fire exposure from Magnusson and Thelandersson (1970) (often referred to as the "Swedish" fire curves). These are derived from heat balance calculations for the burning rate of ventilation-controlled fires. Very different curves are predicted for different ventilation factors and fuel loads. In a similar approach, Lie (2002) performed heat balance calculations for post-flashover fires with a range of ventilation factors and different wall lining materials. Law (1983) used the results from a large number of small-scale tests to propose a simple design method for predicting fire duration and temperatures. The Eurocode (EC1 2002) gives an equation for "parametric" fires, allowing a time–temperature relationship to be produced for any combination of fuel load, ventilation openings, and wall lining materials to give an approximation to the Swedish curves.

These and other approaches have been assessed and compared in the recent engineering guide from SFPE (2004), which recommends the use of the Law method for most applications, with the Magnusson and Thelandersson method and the Lie method recommended in certain cases.

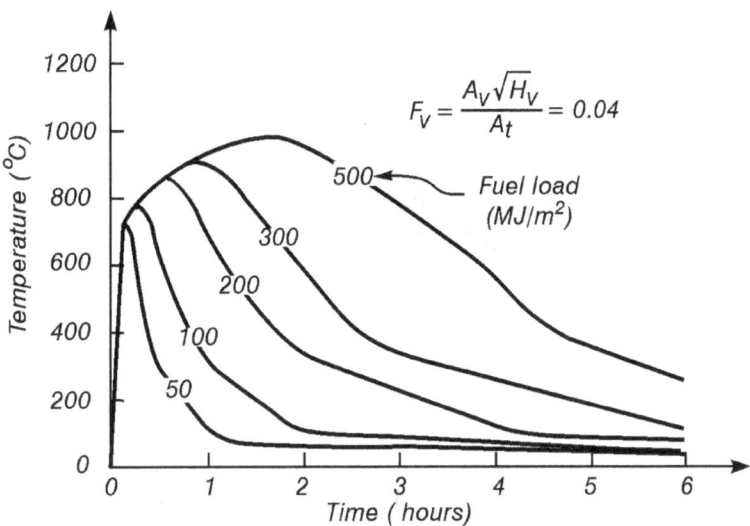

FIGURE 2.3. Typical Time–Temperature Curves for Given Ventilation Factor and Different Fuel Loads (MJ/m^2 of Total Internal Surface Area) (Magnusson and Thelandersson 1970)

Computer programs for calculating temperatures in post-flashover room fires using single-zone models include COMPF2 (Babrauskas 1979), Ozone (Franssen et al. 1999), FASTLite (Buchanan 1997), and CFIRE (Yii 2003).

2.5.4 Time Equivalence

The concept of *equivalent fire severity* is used to relate the severity of an expected real fire to the standard test fire. This is important when designers want to use published fire resistance ratings from standard tests with estimates of real fire exposure. There are several methods of comparing real fires to the standard test fire, the most common being the time equivalence formula given in Eurocode 1 (EC1 2002), which gives the equivalent time t_e (min) as

$$t_e = k_b \, w \, e_f$$

Where:

e_f = Fuel load (MJ/m^2 of floor area)

k_b = A parameter to account for different compartment linings, generally 0.07 min m^2/MJ

w = Ventilation factor, given by

$$w = \left(\frac{6.0}{H_r}\right)^{0.3}\left[0.62 + \frac{90(0.4 - \alpha_v)^4}{1 + b_v \alpha_h}\right] > 0.5$$

H_r = Compartment height (m)

α_v = A_v / A_f $0.05 \leq \alpha_v \leq 0.25$

α_h = A_h / A_f $\alpha_h \leq 0.20$

b_v = $12.5 (1 + 10\,\alpha_v - \alpha_v^2)$

A_f = Floor area of the compartment (m^2)

A_v = Area of vertical openings in the walls (m^2)

A_h = Area of horizontal openings in the roof (m^2)

The equivalent fire severity is very useful where the details of the compartment are known and where the designer wishes to use published fire resistance ratings for selection of construction elements.

2.6 FIRE RESISTANCE OF ELEMENTS EXPOSED TO THE STANDARD FIRE

Fire resistance is a measure of the ability of a building element to resist a fire, usually the time for which the element can meet certain criteria during exposure to a standard fire resistance test. A building element is a structural member such as a beam or a column, a non-structural element such as a partition or door, or a combination such as a floor or load-bearing wall. Individual <u>materials</u> do not possess fire resistance. Fire resistance is a property assigned to <u>building elements</u> that are constructed from a single material or a mixture of materials. A *fire resistance rating* is the fire resistance assigned to a building element on the basis of a test or some other approval system. Some countries use the terms *fire rating, fire endurance rating,* or *fire resistance level,* which are usually interchangeable.

2.6.1 Failure Criteria

The three failure criteria for fire resistance are *stability*, *integrity*, and *insulation*. To meet the *stability* criterion in a standard fire resistance test, a structural element must perform its load-bearing function and carry the applied loads for the duration of the test without structural collapse. The integrity and insulation criteria are intended to test the ability of a barrier to contain a fire, to prevent fire spreading from the room of origin. To meet the *integrity* criterion, the test specimen must not develop any cracks or fissures that allow flame or hot gases to pass through the assembly. To meet the *insulation* criterion, the temperature of the cold side of the test specimen must not exceed a specified limit, usually an average increase of 140°C and a maximum increase of 180°C at a single point (ASTM 2007).

An increasing international trend is for fire codes to specify the required fire resistance separately for stability, integrity, and insulation. For example, a typical load-bearing wall may have a specified fire resistance rating of 60/60/60, which means that a 1-hour rating is required for stability, integrity and insulation, respectively. If the wall was non-load-bearing, the specified fire resistance rating would be - /60/60. A fire door with a glazed panel may have a specified rating of - /30/ - , which means that this assembly requires an integrity rating of 30 minutes, with no requirement for stability or insulation.

2.6.2 Approvals

Most countries require that fire resistance tests be certified by a recognized testing laboratory or approvals agency. In North America, independent testing organizations such as Underwriters Laboratories (UL 2004) and the Southwest Research Institute (SWRI 2004) maintain registers of fire resistance ratings. Most of these ratings are based on standard tests. Ratings based on these approvals are listed in some national building codes (e.g., NBCC 1995, ICC 2006). Some countries may need to use approvals from other countries, so that in New Zealand, for example, a register of approved listings is maintained by the national standards organization (SNZ 1991). Some trade organizations (e.g., ASFPCM 1988, Gypsum Association 2003) maintain industry listings of approvals for products manufactured or used by their members. Listings generally fall into three categories: *generic ratings*, *proprietary ratings*, or *calculation methods*.

Generic fire resistance ratings, or "tabular ratings," are listings that assign fire resistance to typical materials such as concrete or steel. Generic ratings are derived from full-scale fire resistance tests carried out over many years and are widely used because they can be applied to commonly available materials in any country. However, generic ratings make no allowance for the size and shape of the fire-exposed member or the level of load.

Proprietary fire resistance ratings apply to proprietary products made by specific manufacturers, so they may be more accurate than generic ratings but cannot be applied to similar products from other manufacturers.

As fire engineering develops, it is becoming feasible to assess fire resistance of structural members and some assemblies by *calculation*. Some listing agencies and national design codes now include approved calculation methods for assessing fire resistance. Calculation methods must be based on full-scale fire resistance test results of similar assemblies. Calculations can be used for predicting insulation and load-bearing response but not integrity.

An increasing number of listed fire resistance ratings are based on *expert opinion*. The opinion will state whether the assembly would be considered likely to pass a test, based on observations of similar successful tests, calculations, and the considered experience of the testing and approving personnel.

2.7 FIRE RESISTANCE OF BUILDINGS EXPOSED TO REAL FIRES

2.7.1 Design of Steel Buildings Exposed to Fire

Whole buildings or significant assemblies in whole buildings cannot be designed economically by the simple methods described above. It becomes necessary to use specialist computer programs for analysis of fire-exposed structures (the "advanced" calculation method in Eurocodes). Such programs will impose deformations on the structure and calculate the total strain in each member resulting from those deformations. The stress-related strain will be calculated, leading to derivation of the internal forces in each member for comparison with the applied loads. The advanced method is essential for any structures with structural redundancies. The calculated fire resistance of an individual member can be very different from the resistance calculated considering the member to be part of a frame or a building.

2.7.2 Multi-story Frame Buildings

In recent years, the fire performance of large-frame structures has been shown in some instances to be better than the fire resistance of the individual structural elements (Moore and Lennon 1997). These observations have been supported by extensive computer analyses, including Franssen, Schleich, and Cajot (1995) who showed that, when axial restraint from thermal expansion of the members is included in the analysis of a frame building, the behavior is different from that of the column and beam analyzed separately.

A large series of full-scale fire tests was carried out between 1994 and 1996 in the Cardington Laboratory of the Building Research Establishment in England. A full-size eight-story steel building was constructed with composite reinforced concrete slabs on exposed metal decking, supported on steel beams with no applied fire protection other than a suspended ceiling in some tests. The steel columns were fire-protected. A number of fire tests were carried out on parts of one floor of the building, resulting in steel beam temperatures up to 1000 °C, leading to deflections up to 600 mm but no collapse and generally no integrity failures (Martin and Moore 1997).

18

The good performance of the floor/beam systems in such buildings has been attributed to a complex interrelated sequence of events, described rather simply as follows (Buchanan 2001):

1. The fire causes heating of the beams and the underside of the slab.
2. The slab and beam deform downwards as a result of thermal bowing.
3. Thermal expansion causes compressive axial restraint forces to develop in the beams.
4. The reaction from the stiff surrounding structure causes the axial restraint forces to become large.
5. The yield strength and modulus of elasticity of the steel reduce steadily.
6. The downward deflections increase rapidly due to the combined effects of the applied loads, thermal bowing, and the high axial compressive forces.
7. The axial restraint forces reduce due to the increased deflections and the reduced modulus of elasticity, limiting the horizontal forces on the surrounding structure.
8. Higher temperatures lead to a further reduction of flexural and axial strength and stiffness.
9. The slab–beam system deforms into a catenary, resisting the applied loads with tensile membrane forces.
10. As the fire decays, the structural members cool down and attempt to shorten in length.
11. High tensile axial forces are induced in the slab, the beam, and the beam connections.

These actions can take place in two or three dimensions, depending on the geometry of the building and the layout of the structure. The large deformations are often accompanied by local buckling of steel members.

Modern computing power has recently made it possible to model the structural response of buildings exposed to fires. Computer modeling has been used to help interpret the behavior of the Cardington building. Some of the studies have found that the building can be modeled using two-dimensional sub-frames rather than the complete three-dimensional frame, but others have emphasized the three-dimensional behavior. The development of tensile membrane action in reinforced concrete or composite steel/concrete floors is described by Lim et al. (2004).

2.8 MATERIALS STANDARDS

In most countries, the materials standards for structural design provide methods of assessing or calculating fire resistance. These often use the three levels shown in Table 2.2.

2.8.1 U.S.A.

Structural design for fire resistance in the United States has not moved to performance-based design as quickly as in Europe. A very useful overview of the effects of fires on structures is given by Lie (1992). Existing building codes include prescriptive requirements for fire resistance, which have not changed greatly in recent years, and the current movement from regional to national building codes (IBC, NFPA codes) has not been accompanied by significant changes in design for fire resistance. However, several background documents have recently been published (e.g., SFPE 2004) that will eventually lead to changes in materials standards.

Guidance on the severity of fire exposure on structural elements for both fully developed fires and for fire plumes is given in SFPE 2004.

The most recent U.S.-based standard for structural fire calculations is the joint ASCE/SFPE Standard 29-05 (ASCE/SFPE 2005), which gives simple calculation methods for all main materials of construction. Most of the methods in this document are empirical methods strongly based on standard fire resistance testing, with very little guidance on sophisticated analysis or design from first principles.

Industry groups for particular materials (steel, concrete, and timber) are also developing standards and guidance documents for structural fire resistance.

2.8.1.1 Steel

The American Institute of Steel Construction (AISC) is the principal organization providing documentation for design of steel structures. The AISC, March 2005 *Specification for Structural Steel Buildings* (ANSI/AISC 350-05) provides for design of steel structures for fire conditions. Specifically, the 2005 Specification addresses design by engineering analysis and provides guidance on load combinations and required strength, design-basis fires and both simple and advanced methods of analysis. Simple methods of analysis are applicable to individual members while advanced methods are applicable to entire steel building frames. Additionally, ANSI/AISC 360-05 gives basic information on thermal and mechanical properties of steel and concrete at elevated temperatures.

The American Institute of Steel Construction (AISC 2003) recently commissioned a major strategy report on integrating the structural engineering and fire engineering of steel structures. The report gives a survey of existing codes and standards, plus background information on fire testing, analysis, and design methods for steel structures.

The *SFPE Handbook of Fire Protection Engineering* (SFPE 2002) has a chapter on steel design that gives an overview of steel structures' performance in fire, but this does not give sufficient information for the advanced calculation methods in Eurocode 3.

The American Iron and Steel Institute has published a manual for load and resistance factor design of cold-formed steel framing members (AISI 1991).

2.8.1.2 Concrete

The American Concrete Institute is the principal organization providing documentation for design of concrete structures. ACI/TMS 216.1-07, *Code Requirements for Determining Fire Resistance of Concrete and Masonry Construction Assemblies,* provides basic information for design of concrete structures to resist standard fire exposure. This document is being updated with new guidance for fire design.

The Concrete Reinforcing Steel Institute has a handbook on fire resistance of reinforced concrete (CRSI 1980), and the *SFPE Handbook of Fire Protection Engineering* (SFPE 2002) has a chapter on concrete design that gives an overview of concrete structures performance in fire.

2.8.2 Canada

Most design standards (concrete, wood, and steel) in Canada refer to the National Building Code of Canada (NBCC 2005) for fire resistance specifications, and a new standard for fiber-reinforced plastics (FRP) is CSA-S806, which was published in 2002 and includes design charts for the fire resistance design of FRP-reinforced concrete slabs.

Guidance for design of concrete structures in fire is given by Harmathy (1993), who draws on Canadian research and experience.

2.8.3 Eurocodes

By far the most comprehensive international documents for structural design of buildings and structures in fire conditions are the Structural Eurocodes. The main codes follow, with details in the list of references in Section 2.10 below:

- EN 1991 Eurocode 1 Basis of design and actions on structures
- EN 1992 Eurocode 2 Design of concrete structures
- EN 1993 Eurocode 3 Design of steel structures
- EN 1994 Eurocode 4 Design of composite steel and concrete structures
- EN 1995 Eurocode 5 Design of timber structures
- EN 1996 Eurocode 6 Design of masonry structures
- EN 1997 Eurocode 7 Geotechnical design
- EN 1998 Eurocode 8 Design provisions for earthquake resistance of structures
- EN 1999 Eurocode 9 Design of aluminum alloy structures

Most of these have substantial fire sections (100 pages or more).

Most European countries also have a national "Annex" to provide various nationally determined parameters and other information, sometimes in the form of a "National Application Document." There will be an opportunity to review the Eurocodes at five-year intervals, and there may be a maintenance group to answer questions or correct any errors in the shorter term.

Most of the Structural Eurocodes include the following statement:

> "A full analytical procedure for structural fire design would take into account the behavior of the structural system at elevated temperatures, the potential heat exposure and the beneficial effects of active and passive fire protection systems, together with the uncertainties associated with these three features and the importance of the structure (consequences of failure).

> "At the present time it is possible to undertake a procedure for determining adequate performance which incorporates some, if not all, of these parameters and to demonstrate that the structure, or its components, will give adequate performance in a real building fire. However, where the procedure is based on a nominal (standard) fire the classification system, which call for specific periods of fire resistance, takes into account (though not explicitly), the features and uncertainties described above."

Design can then be at various levels in a hierarchy, as shown in Figure 2.4, which identifies both the prescriptive and the performance-based approaches. In general, the simplest designs will be at the far left-hand side of Figure 2.4 (tabulated data for single members in a prescriptive environment), with the most sophisticated designs being at the right-hand side of Figure 2.4 (advanced calculation models for entire structures).

All the Structural Eurocodes include the following sections:

- Basis of design
    - Fire exposure
    - Verification methods
    - Methods of structural analysis
- Material properties
    - Mechanical properties
    - Thermal properties
- Design procedures
    - Tabulated data
    - Simple calculation methods
    - Advanced calculation methods
- Construction details

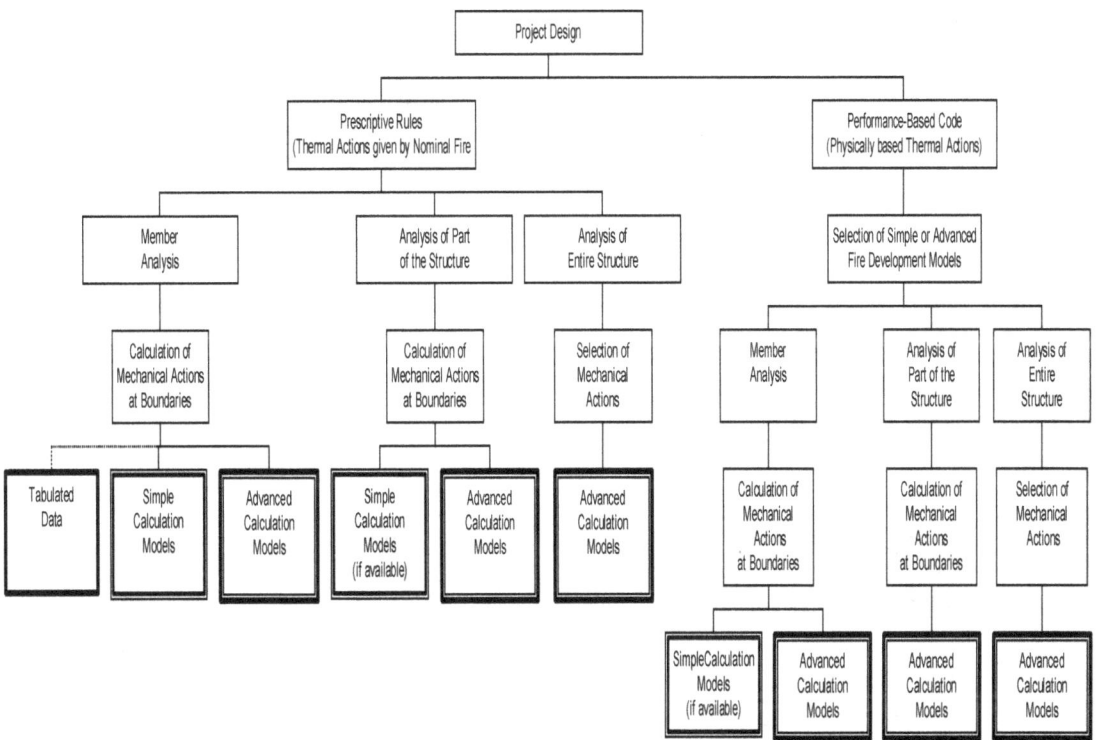

FIGURE 2.4. Alternative Design Procedures in the Structural Eurocodes (EC2 2002)

The fire exposure allows for standard or realistic fire design curves to be used. The *simple calculation methods* are for predicting the behavior of single members based on simple assumptions. The *advanced calculation methods* provide the principles for computer analyses based on fundamental physical behavior for both thermal analysis and mechanical behavior. These analyses need to take into account factors such as transient temperature gradients, variation of thermal properties with temperature, axial and flexural restraint, thermally induced forces, and thermally induced deformations, throughout the duration of the expected fire. The effects of creep are not explicitly included in the advanced calculation methods, but the stress–strain relationships have been modified to include creep in an indirect way.

The Eurocodes include information that does not generally appear in other fire codes, such as comprehensive expressions for thermal and mechanical properties at elevated temperatures and stress-strain relationships at elevated temperatures. This is very useful for any analytical modeling of fire behavior of structures. The tabulated listings in the Eurocodes are far more extensive than most other codes, the particular benefit to designers being that the tables include the improved fire resistance for members that are loaded below their design strength at the time of a fire.

2.8.4 European Countries

All the major European countries have been involved in development of the Eurocodes, but they have also been maintaining parallel development of national codes that are used for everyday design. The transition to design office use of the Eurocodes is expected to be slow in most countries, depending on the rate at which the existing national codes are phased out.

In the United Kingdom, a comprehensive recent publication, *Structural Response and Fire Spread Beyond the Enclosure of Origin* (BSI 2003), is a "Published Document" in support of BS 7974, *Application of Fire Safety Engineering Principles to the Design of Buildings* (BSI 2001). BS 7974 is currently the most comprehensive code of practice for specific fire engineering design in any country.

The U.K. Published Document (150 pages) is complementary to the Structural Eurocodes and provides data and guidance for calculating the fire exposure and fire resistance (structural and non-structural) for a wide range of materials and assemblies. The document recognizes that detailed structural analysis of complex load-bearing structural frames is beyond the scope of such a guidance document. More detail is given in specific material codes such as the *Code of Practice for Design of Structural Steelwork* (BSI 1990).

2.8.5 Australia and New Zealand

The Australian and New Zealand fire codes permit specific fire engineering design in a similar performance-based environment. However, the minimum fire ratings specified by the New Zealand prescriptive documents are much lower than in Australia and many other countries because of more emphasis on life safety than property protection. The fire requirements in the structural design codes are rather simplistic, using tabulated values for reinforced concrete, for example, or specifying that standard tests should be used for establishing fire resistance ratings. All fire resistance values are based on standard fire exposure with little or no mention of realistic fires. Alternative calculations are permitted but, unlike Europe, very little guidance is given (Buchanan 2000). A useful *Guide for the Design of Fire Resistant Barriers and Structures* has recently been published in Australia (England et al. 2000).

2.9 CONCLUSIONS

- The international community is moving toward performance-based engineering standards for structural design in fire conditions.
- The Structural Eurocodes provide the most comprehensive current codified source of information on design for structural fire resistance.
- There is a developing spectrum of fire design methods ranging from simple tabulated data to advanced structural analysis and design techniques.

- The more simple the design method, the more conservative the underlying assumptions need to be in order to provide the desired level of protection against collapse in fire.
- A major limitation on structural design for fire conditions is assessment of the fire scenario and the resulting fire temperatures.
- Advanced structural analysis and design of buildings in fire conditions are more difficult than for normal temperature conditions.

2.10 REFERENCES

ACI (2007), "Code Requirements for Determining Fire Resistance of Concrete and Masonry Construction Assemblies," ACI/TMS 216.1-07, Farmington Hills, Mich.: American Concrete Institute.

AISC (2005), *Steel Construction Manual,* 13th ed., Chicago: American Institute of Steel Construction.

AISC (2003), "Strategy for Integrating Structural and Fire Engineering of Steel Structures," Report prepared by Ove Arup & Partners for American Institute of Steel Construction.

AISI (1991), *Load and Resistance Factor Design Specification for Cold-Formed Steel Structural Members*, Washington: American Iron and Steel Institute.

ASCE (2005), *Minimum Design Loads for Buildings and Other Structures*, SEI/ASCE 7-05, New York: American Society for Civil Engineers.

ASCE/SFPE (2005), *Standard Calculation Methods for Structural Fire Protection*, ASCE/SEI/SFPE 29-05, Reston, VA: American Society of Civil Engineers.

ASFPCM (1988), *Fire Protection for Structural Steel in Buildings*, U.K.: Association of Specialist Fire Protection Contractors and Manufacturers.

ASTM (2007), *Standard Test Methods for Fire Tests of Building Construction and Materials*, E119-07a, West Conshohocken, Pa.: American Society for Testing and Materials.

Babrauskas, V. (1979), *COMPF2: A Program for Calculating Post-Flashover Fire Temperatures*, NBS Technical Note 991, National Bureau of Standards.

BSI (1987), *Fire Tests on Building Materials and Structures*, BS 476 (Parts 1 to 23), U.K.: British Standards Institution.

BSI (1990), "Structural Steelwork for Use in Building" – *Part 8: Code of Practice for Fire Design*, BS 5950-8, U.K.: British Standards Institution.

BSI (2001), *Application of Fire Safety Engineering Principles to the Design of Buildings - Code of Practice*, BS 7974:2001, U.K.: British Standards Institution.

BSI (2003), *Application of Fire Safety Engineering Principles to the Design of Buildings - Part 3: Structural Response and Fire Spread Beyond the Enclosure of Origin (Subsystem 3)*, Published Document PD 7974-3, U.K.: British Standards Institution.

Buchanan, A.H. (1994), "Fire Engineering for a Performance Based Code, *Fire Safety Journal*. 23:1 (1994) 1-16.

Buchanan, A.H. (1997), "Modeling Post Flashover Fires with FASTLite," *Journal of Fire Protection Engineering* 9:3 (1997) 1-11.

Buchanan, A.H., ed. (2000), *Fire Engineering Design Guide*, New Zealand: Centre for Advanced Engineering, University of Canterbury.

Buchanan, A.H. (2001), *Structural Design for Fire Safety*, U.K.: John Wiley & Sons, Ltd.

CRSI (1980), *Reinforced Concrete Fire Resistance*, Chicago: Concrete Reinforcing Steel Institute.

Custer, R.L.P., and B.J. Meacham (1997), *Introduction to Performance-Based Fire Safety*, Society of Fire Protection Engineers and National Fire Protection Association.

EC1 (2002), Eurocode 1: *Actions on Structures. ENV 1991, Part 1-2: General Actions—Actions on Structures Exposed to Fire*, Brussels: European Committee for Standardization.

EC2 (2002), Eurocode 2: *Design of Concrete Structures. ENV 1992, Part 1-2: General Rules—Structural Fire Design*, Brussels: European Committee for Standardization.

EC3 (2002), Eurocode 3: *Design of Steel Structures. ENV 1993, Part 1-2: General Rules—Structural Fire Design*, Brussels: European Committee for Standardization.

EC4 (2003), Eurocode 4: *Design of Composite Steel and Concrete Structures. ENV 1994, Part 1-2: General Rules—Structural Fire Design*, Brussels: European Committee for Standardization.

EC5 (1994), Eurocode 5: *Design of Timber Structures. ENV 1995, Part 1-2: General Rules—Structural Fire Design*, Brussels: European Committee for Standardization.

EC6 (1995), Eurocode 6: *Design of Masonry Structures. ENV 1996, Part 1-2: General Rules—Structural Fire Design*, Brussels: European Committee for Standardization.

Ellingwood, B.R., and R.B. Corotis (1991), "Load Combinations for Buildings Exposed to Fires," *Engineering Journal*, American Institute of Steel Construction 28:1 (1991) 37-44.

England, J.P., S.A. Young, M.C. Hui, and N. Kurban (2000), *Guide for the Design of Fire Resistant Barriers and Structures.* Melbourne: Building Control Commission.

Feasey, R., and A.H. Buchanan (2002), "Post-flashover Fires for Structural Design," *Fire Safety Journal* 37 (2002) 83-105.

Franssen, J-M., J-B. Schleich, and L-G. Cajot (1995), "A Simple Model for the Fire Resistance of Axially Loaded Members According to Eurocode 3," *Journal of Constructional Steel Research* 35 (1995) 49-69.

Franssen, J.-M., V.K.R. Kodur, and J. Mason (2002), *User's manual for SAFIR2001 free: A computer program for analysis of structures at elevated temperature conditions*, Liege, Belgium: University of Liege.

Franssen, J-M., et al. (1999), *Competitive Steel Buildings Through Natural Fire Safety Concept*, Draft Final Report, Part 2, Luxembourg: Profil Arbed Centre de Recherches.

Gypsum Association (2003), *Fire Resistance Design Manual*, 17th ed., Washington: Gypsum Association.

Harmathy,T.Z. (1993), *Fire Safety Design and Concrete*, Essex, England: Longman Scientific & Technical.

ICC (2006), *International Building Code*, Falls Church, Va.: International Code Council.

ICC (2003b), *Performance-Based Building Design Concepts: A Companion Document to the ICC PC*, Falls Church, Va.: International Code Council.

ICC (2006), *ICC Performance Code for Buildings and Facilities*, Falls Church, Va.: International Code Council.

ISO (1975), *Fire Resistance Tests—Elements of Building Construction*, ISO 834 – 1975, International Organization for Standardization.

Law, M. (1983), "A Basis for the Design of Fire Protection of Building Structures," *The Structural Engineer* 61A:5 (1983).

Lie, T.T. (1992) (ed.), "Structural Fire Protection," ASCE Manuals and Reports of Engineering Practice No 78, New York: American Society of Civil Engineers.

Lie, T.T. (2002), "Fire Temperature-Time Relations," Chapter 4-8, *SFPE Handbook of Fire Protection Engineering*, 3rd ed., Bethesda, Md.: Society of Fire Protection Engineers.

Lim, L., A.H. Buchanan, P.J. Moss, and J-M. Franssen (2004), "Computer Modeling of Restrained Reinforced Concrete Slabs in Fire Conditions," *ASCE Journal of Structural Engineering* 130:12 (December 2004) 1964-1971.

Magnusson, S.E., and S. Thelandersson (1970), *Temperature-Time Curves of Complete Process of Fire Development: Theoretical Study of Wood Fuel Fires in Enclosed Spaces*, Acta Polytechnica Scandinavica, Civil Engineering and Building Construction Series 65.

Martin, D.M., and D.B. Moore (1997), Introduction and Background to the Research Programme and Major Fire Tests at BRE Cardington, Proceedings, London: National Steel Construction Conference.

Moore, D.B., and T. Lennon (1997), "Fire Engineering Design of Steel Structures," *Progress in Structural Engineering and Materials* 1:1 (1997) 4-9.

NBCC (1995), *National Building Code of Canada*, Ottawa: National Research Council of Canada.

NFPA (2006), "Standard Methods of Tests of Fire Endurance of Building Construction and Materials," NFPA 251, Quincy, Mass.: National Fire Protection Association.

NFPA (2006), *Building Construction and Safety Code, NFPA 5000*, Quincy, Mass.: National Fire Protection Association.

SAA (1990), *Fire Resistance Tests of Elements of Structure*, AS 1530.4-1990, Standards Association of Australia.

SFPE (2000), *Engineering Guide: Performance-Based Fire Protection*, Bethesda, Md.: Society of Fire Protection Engineers.

SFPE (2002), *SFPE Handbook of Fire Protection Engineering*, Bethesda, Md.: Society of Fire Protection Engineers.

SFPE (2004), *Engineering Guide: Fire Exposures to Structural Elements*, Bethesda, Md.: Society of Fire Protection Engineers.

SNZ (1991), *Fire Properties of Building Materials and Elements of Structure,*. Miscellaneous Publication No. 9, Wellington: Standards New Zealand.

SWRI (2004), *Directory of Listed Products*, San Antonio, Texas: Southwest Research Institute, Department of Fire Technology.

UL (2003), *Fire Tests of Building Construction and Materials*, UL 263, Northbrook, Ill.: Underwriters Laboratories Inc.

UL (2004), *Fire Resistance Directory*, Northbrook, Ill.: Underwriters Laboratories Inc.

ULC (2004), *Standard Methods of Fire Endurance Tests of Building Construction and Materials*, CAN/ULC-S101-04, Toronto: Underwriters Laboratories of Canada.

Yii (2003), "Modelling the Effects of Fuel Types and Ventilation Openings on Post-Flashover Compartment Fires," Fire Engineering Research Report 03/1, New Zealand: University of Canterbury.

Chapter 3

Decision Framework for Fire Risk Mitigation
Bruce R. Ellingwood, Ph.D., P.E., Georgia Institute of Technology

3.1 INTRODUCTION

The public generally has been well served by building code provisions for fire protection and fire safety. Since the early 20th century, code requirements for passive fire protection traditionally have been derived from building component qualification testing (according to ASTM Standard E119, in the United States). Prescriptive code design requirements and methods (e.g., *2003 International Building Code*, Sections 703 and 720; *NFPA Building Construction and Safety Code*, NFPA 5000, Section 8.2.2) stipulate acceptance criteria (in the form of fire ratings) that are based on a component surviving a "standard" fire for a prescribed rating period. From a practical viewpoint, the traditional approach has proved relatively easy to implement and to codify and enforce through building regulation. Fire ratings can be useful for classification purposes or for making comparisons of performance of structural components and other building products under standardized conditions as well as for demonstrating code compliance. On the other hand, such prescriptive requirements and ratings often are simplistic and seldom are indicative of actual building performance during a fire. They are based on experience and are usually, but not always, conservative. They stipulate an unrealistic fire (one that presumes an inexhaustible fuel supply during the rating period), do not distinguish differences in compartment ventilation or surface composition, and do not account for realistic structural loads, thermal effects, or conditions of structural restraint. They do not account for innovations that have taken place in modern building construction. Perhaps most importantly, they focus on fires that are localized in compartments (implied by the requirement that floors and walls are to be heated on only one side) and do not address the impact of the fire on the structural system as a whole. As a result, current fire protection practices may lead to inefficient, uneconomical, and occasionally inadequate design solutions. The performance of building systems during realistic fires often is better than anticipated (Buchanan 2001). Despite this, building structural components and systems that are known to perform acceptably under realistic fire exposures may be penalized or not permitted by current practices (Milke 1985; Meacham 1997; Kruppa 2000; Bennetts and Thomas 2002). Finally, the specific performance objectives of most prescriptive fire protection provisions are not well articulated, making it difficult to apply them to non-routine design situations.

The new paradigm of performance-based engineering, where there is a strong motivation to seek alternatives to ratings based on prescriptive design and qualification testing, is moving the building design profession in this direction in the area of fire safety assurance. Performance-based engineering enhances the prospect of clarifying the intent of the code, overcoming the need to rely on prescriptive design solutions that may be disconnected from reality, and providing a framework for innovative design solutions. In contrast to traditional prescriptive approaches to fire protection, performance-based fire engineering (PBFE) requires a systematic approach to identifying building performance objectives and quantitative structural analysis tools to verify that these objectives have been achieved. In the United States, performance-based engineering solutions for fire protection are permitted under the "alternate means and methods" or "equivalency" provisions of model building codes (e.g., Section 1.5 of NFPA 5000; Section 104.11 of the IBC; and the performance-based design option in Chapter 5 of NFPA 5000). However, the lack of technical methods and data has inhibited their implementation for all but special buildings. With recent advances in fire science and advanced structural analysis as a design tool, it is becoming possible to consider realistic fire scenarios and fire effects on the building's structural system as a whole as part of the design process (Buchanan 2001; SFPE 2000).

In an era when prescriptive building code requirements for fire protection were the norm, structural engineers seldom were responsible for fire protection of building structural systems. Such protection has been mainly the responsibility of the project architect and, occasionally, a fire protection engineer. This state of affairs is changing with the move toward PBFE. There are a number of instances where structural engineering for fire conditions may add value to the building design process. One case in point is when innovative architectural expressions can be inhibited by customary fire resistance rating requirements (Siu 2004). Another is by providing a level of performance for buildings of unique social or economic importance beyond the traditional goals of safeguarding lives and property, for example in protecting irreplaceable building contents against the consequence of building failure. In some cases, a building owner may wish to assess general structural integrity and the likelihood of progressive collapse if fire protection has been removed by the effect of other abnormal loads (Liew and Chen 2004). Finally, nonconforming fire code issues with existing construction can be addressed efficiently with PBFE prior to undertaking costly rehabilitation.

A fire with the potential to damage a building structure severely is a low-probability event in comparison with the events that give rise to other loads and structural actions that are common to structural engineering analysis and design. Severe fires can lead to ultimate structural limit states such as gross inelastic deformation, instability, or partial or total building collapse. The science of fire-resistant structural design is at an early stage of development, and the structural engineering profession lacks the customary engineering tools to attack the problem. Many building codes and standards such as ASCE Standard 7-05 (ASCE 2005) contain a requirement to provide general structural integrity, which is aimed at mitigating events that are outside the design envelope. These provisions generally lack specifics, however, and structural engineers find them difficult to apply.

Most factors that determine building safety under fire conditions are uncertain in nature. In the presence of uncertainty, no building system can be engineered and constructed to be absolutely risk free from the effects of fire. Rather, the fire risk must be managed by a combination of measures involving architectural and structural engineering, building systems engineering, and occupant education. With their extensive provisions for fire safety, building codes are (and have been) key tools for managing fire risk in building construction in the interest of public safety, but the risks addressed by code provisions have been managed judgmentally. The aftermath of recent natural and man-made disasters has necessitated a re-evaluation of such judgmental approaches to risk management. Mitigation of risk from low-probability/high-consequence events, such as fire, through the added dimension of structural analysis and design requires a different approach from the one taken in present building codes. Questions regarding alternative or innovative fire protection strategies and required structural strength to withstand a severe fire can only be answered from a risk perspective. The move toward PBFE will require risk-informed assessments of fire hazards and alternative strategies for hazard mitigation.

This chapter introduces basic concepts of modern risk-informed decision making and suggests a framework for developing and implementing structural design requirements for mitigating fire risk in the current building regulatory climate.

3.2 PERFORMANCE OBJECTIVES AND RISK ASSESSMENT FOR NATURAL AND MAN-MADE HAZARDS

Performance-based fire engineering requires quantitative goal setting and documentation on the part of the design team. Risk measures are important in PBFE since they become the basis for measuring compliance with performance objectives, for comparing alternatives, and for highlighting the role of uncertainty in the decision process.

3.2.1 Performance-Based Engineering

Performance-based engineering (PBE) is evolving to enable new building technologies and structural design to better meet heightened public expectations and to enable more reliable prediction and control of building performance. The perception that some structural systems designed to code by current building practices have failed to perform satisfactorily during recent natural and man-made disasters certainly has provided impetus for the move toward PBE, but the desire to minimize arbitrariness and add value to the building process also is a significant motivator. Efforts in this regard are well under way in Europe, Australia, New Zealand, Canada (where the expected PBE work product for the 2005 *National Building Code of Canada* is termed an "objective-based code"), and the United States (ICC 2003; NFPA 2002). Performance-based engineering is based on:

- A hierarchical set of explicitly stated functional requirements related to building category and hazard intensity
- Quantitative criteria to ensure minimum attributes (e.g., strength, stiffness, durability) necessary to meet those requirements
- Evaluation methods (analysis or test) by which satisfaction of the criteria can be measured
- Extensive commentaries to explain the basis of the criteria and evaluation methods and to provide guidance in their application (Hamburger 1996; Ellingwood 1998)

To date, PBE has focused particularly on two areas: fire engineering and earthquake engineering. The motivating factors behind PBE in the fire engineering area are strongly economic in nature. The Society of Fire Protection Engineers (SFPE) is moving its standards program for fire-resistant design toward PBFE (SFPE 2000), and the American Institute of Steel Construction (AISC) has developed a new Appendix 4 on structural design for fire conditions. The International Council for Research and Innovation in Building and Construction (CIB) has ongoing work on performance-based fire-safety design through its Working Commission W14 (1983; Thomas et al. 1986; 2001), and the European Convention for Constructional Steelwork (ECCS) has developed a model performance-based fire engineering code (ECCS 2001). Such activities on the international scene will accelerate the development of improved quantitative methods for engineering structures for fire safety.

Performance objectives for fire-resistant design now appear in the two Model Codes (ICC 2006; NFPA 2006). For example, Section 1701.3.11 of the *ICC Performance Code for Buildings and Facilities* (ICC 2006) states that:

> "Structural members and assemblies shall have a fire resistance appropriate to their function, the fire load, the predicted fire intensity and duration, the fire hazard, the height and use of the building, the proximity to other properties, and any fire protection features."

Sections 5.2.2 and 5.2.3 of NFPA's *Building Construction and Safety Code* (NFPA 5000) contain similar statements. The intent of such statements seems clear, but the wording is vague, and many engineers (and code officials) are unwilling to undertake the responsibility (liability) associated with implementing the concept in design.

Most proposals for PBE have included a performance matrix in which one axis describes event severity (e.g., small, medium, large) or frequency, while the second axis identifies building occupancy classifications such as those that appear in Table 1-1 of ASCE Standard 7-05 (ASCE 2005). A common factor in these proposals is a multi-tiered approach to design. The stipulated event impacts (e.g., minor, moderate, severe) are placed at the intersections of rows and columns in the matrix to define the level of performance expected for each occupancy classification. An illustration of such a matrix developed for the *ICC Performance Code* (ICC 2006) is provided in Table 3.1. Such matrices have been developed mainly for natural hazards—the earthquake hazard, in particular. Interpreting Figure 3.1, one might expect that a "large" fire would have a severe impact on Category I occupancies (temporary or storage facilities with low hazard to human life), a high impact on Category II occupancies (office or multi-story residential buildings) and a moderate impact on Category III facilities (e.g., hospitals, schools). As an alternative, it might be observed that Table 3.1 could be re-arranged to display severity (or frequency) of the event on the vertical axis and consequence on the horizontal axis, placing building occupancy categories at the intersections of rows and columns in the matrix. (Such a format was followed in an early proposal by the Structural Engineers Association of California for performance-based earthquake design.) While such performance matrices are useful for

TABLE 3.1. Building Performance Matrix (ICC 2006)

"Size" of event	Perf. Group I	Perf. Group II	Perf. Group III	Perf. Group IV
V. Large	Severe	Severe	High	Mod
Large	Severe	High	Mod	Mild
Medium	High	Mod	Mild	Mild
Small	Mod	Mild	Mild	Mild

organizing the design process, their applicability to PBFE bears further examination because, in contrast to earthquake-related performance, the building performance during fire evolves during the event depending on the fire suppression systems that are called into play.

Risk analysis tools are essential to the success of PBFE for measuring compliance with performance objectives, for comparing alternatives, and for highlighting the role of uncertainty in the decision process (Ellingwood 2005b).

3.2.2 Risk and Its Analysis—Hazard, Consequences, Context

Risk involves *hazard, consequences,* and *context* (Elms 1992; Stewart and Melchers 1997). The hazard is a potentially harmful event, action, or state of nature. The potential for a fire in a building is a hazard. The occurrence of the hazardous event has consequences—building damage or collapse, loss of life or personal injury, economic losses, or damage to the environment—that must be measured in some manner, as described below. Finally, there is the context, which provides a frame of reference for the risk analysis, assessment, and decision. As stakeholders in a building risk assessment and decision process, individuals, management groups, government agencies, or other decision makers may view risk differently. Individuals seldom undertake risky activities without an expectation of some benefit. Most individuals or small groups are risk averse (implying that they require a substantial increase in value or benefit in return for accepting marginal increases in risk). On the other hand, governments and large corporations, which may be self-insured, tend to be risk neutral. Willingness on the part of individuals to accept risk depends on whether the risk is undertaken voluntarily or involuntarily (Starr 1969) and whether the individual *perceives* that he or she can manage the risky situation. Incidents involving large numbers people are viewed differently from incidents involving individuals. The element of familiarity or dread or the unknown in *perception of risk* plays a significant role in risk acceptance. The context also is determined by the necessity for risk management and how additional investment in risk reduction is balanced against available resources.

Building codes and structural design practice aim at delivering building products and systems with risks that the public finds acceptable. Despite the advances in structural reliability analysis and acceptance of probability-based structural codes (Galambos et al. 1982; Ellingwood et al. 1982), it remains unclear exactly what is *acceptable risk* in the built environment. Like other risks, acceptable risk in building construction is relative in the sense that it can be determined only in the context of:

- What is acceptable in other activities
- What investment is required to marginally reduce the risk
- What losses might be incurred if the risk were to increase

To a building occupant, any risk below the (unknown) threshold is acceptable. To a developer, on the other hand, any risk above the threshold represents wasted cost.

Annual mortality statistics in the United States provide a psychological yardstick, of sorts, in measuring and discussing risk in terms of annual frequency, although these risks are not truly comparable to building risk. For the healthy adult population, the mortality risk from cardiovascular disease and cancer is on the order of 10^{-3}/year. At the other extreme, the *de minimis* risk, that risk below which society normally does not impose any regulatory guidance, is on the order of 10^{-7}/year (Pate-Cornell 1994). Between these annual frequencies of 10^{-3}/year and 10^{-7}/year is a gray area in which measures to reduce risk usually are traded off against increments in cost of risk reduction. For the sake of illustrating the role of risk in PBFE in this chapter, we may take 10^{-6}/year as the upper threshold of acceptable risk (measured in terms of annual frequency) due to fires in building construction. In terms of order of magnitude, this is not inconsistent with the failure probability of building systems from other natural events (Ellingwood 2001). In first-generation LRFD (load and resistance factor design) (Galambos et al. 1982; Ellingwood et al. 1982), the target member limit state probability involving formation of the first plastic hinge was approximately 0.001 in 50 years (corresponding to a "reliability index" of about 3.0); annualized, this is on the order of 10^{-5}. The annual probability of partial or total collapse of a redundant structural frame is approximately one order of magnitude less, or on the order of 10^{-6}/yr. Note that such comparisons assume that "risk" is equivalent to "annual probability or frequency."

It would be tempting to assert that the acceptable risk for PBFE should be set so that the PBFE design alternatives are at least as safe as those that comply with existing prescriptive requirements. This line of thinking is analogous to that followed in first-generation LRFD, which was calibrated (in an overall reliability sense) to existing structural design practice. When applied to PBFE, however, this approach is questionable. The calibration process for structural components subjected to dead, live, wind, and snow load drew upon years of successful experience in designing for those common loads using recognized principles of structural mechanics and behavior. In contrast, fire is a low-probability event; moreover, the current fire-resistant design approach cannot be tied in any meaningful experiential way to real structural demands, behavior, or response. May (2004) has argued that, rather than to argue about what is acceptable risk, performance-based engineering should aim at developing tools that would allow a stakeholder or decision maker to make informed choices about how to manage the risk.

Building risk must be measured quantitatively to be useful in decision making. The risk metric can be expressed as a probability of failure to meet a performance objective. That probability can be evaluated from the following equation:

$$P[\text{Loss} > \theta\ ] = \Sigma_H \Sigma_{LS} \Sigma_D\ P[\text{Loss} > \theta|D]\ P[D|LS]\ P[LS|H]\ P[H] \tag{3.1}$$

Where:

$P[A]$	=	Probability of event A
$P[A|B]$	=	(Conditional) probability of event A, given the occurrence of event B
θ	=	An appropriate loss metric: severe injury or death, direct damage costs, loss of opportunity costs, etc.
$P[LS|H]$	=	Conditional probability of a structural limit state
$P[D|LS]$	=	Conditional probability of damage state (e.g., negligible, minor, moderate, severe)
$P[\text{Loss} > \theta|D]$	=	Conditional probability of loss

The probability, P[H], defines the hazard probabilistically. In applications familiar to many structural engineers (e.g., wind, earthquake, and flood hazard), it often is expressed as a function of the intensity of the event (wind speed, spectral acceleration, or flood stage) and in that form is termed the "hazard curve."

As an alternative, the risk assessment may be based on a set of stipulated scenario events rather than on a hazard with a random intensity, depending on the preferences of the decision-maker. Each scenario represents a description of fire development in time, from ignition through full development and decay, with key features identified that distinguish it from other fires. For a scenario fire, the risk metric in Eq. 3.1 becomes a conditional probability:

$$P[\text{Loss} > \theta | H_s] = \Sigma_{LS} \Sigma_D \, P[\text{Loss} > \theta | D] \, P[D|LS] \, P[LS|H_s] \tag{3.2}$$

in which H_s = scenario event(s) selected. Specific interpretations of Eqs. 3.1 and 3.2 for fire-resistant design are discussed below in Section 3.3.

Equations 3.1 and 3.2 deconstruct the risk analysis into its major constituents and enable the design team and decision makers (representing different technical disciplines) to focus on strategies where risk mitigation is most likely to be achieved successfully and economically. The likelihood of the hazard is measured by P[H] (or by its mean annual frequency, λ_H; as noted below, the two are indistinguishable for rare events). The probabilities P[LS|H] or P[LS|H_s] are determined by structural engineering analysis. The P[D|LS] describes the damage state for the structural system in terms of the structural response quantities computed from the structural analysis. Finally, the conditional probability, P[Loss > θ|D], describes the probability of loss, given a specific damage state. Both approaches have their advantages and drawbacks in specific situations. Eq. 3.1 can be used to assess risk for a spectrum of events and to estimate losses over time (often on an annual basis, as that is the common way of reporting λ_H and resulting losses). On the other hand, it may not be practical to identify and/or analyze the full spectrum of hazards. Furthermore, the occurrence of some hazards, such as deliberate fires or other acts of malevolence directed at specific targets, cannot be modeled probabilistically at the current state of the art of risk analysis. Scenario analysis (Eq. 3.2) usually considers a relatively small number of hazardous situations, each of which may be described in considerable detail (Hurley 2004). This approach allows the decision makers to focus on events that are deemed to be particularly significant to building performance and facilitates communication of the design-basis events to the building stakeholders. However, the probability of each scenario seldom is calculated. Thus, the probabilities in Eq. 3.2 are conditional in nature, and as a consequence the loss probabilities cannot be annualized or benchmarked against other commonplace risks.

A key ingredient of risk management is identification of appropriate risk metrics for the event {Loss > θ} in Eqs. 3.1 and 3.2. For purposes of this chapter, this event should be interpreted in a broad sense. Structural codes traditionally have been concerned first and foremost with public safety (preventing loss of life or personal injury for a normative set of design hazards) and property protection, and in this context the collapse of a building, or a large portion of it, is a surrogate for all other metrics. Occupant and public safety will continue to be the primary objectives of building codes in performance-based engineering. Other performance metrics—direct economic losses from building contents damage, indirect losses due to interruption of function, foregone opportunities, and loss of amenity—traditionally have not been addressed by the building regulatory community but may be of concern to certain stakeholder groups for certain types of building facilities. Some in the building community have voiced the opinion that an appropriate performance objective for fire-resistant structural design is for the building system to survive burnout without structural collapse. The purpose here is not to judge the merits of these alternative performance objectives (that is a major ingredient of the goal-setting in PBFE that must occur among stakeholders), but rather to note that Eqs. 3.1 and 3.2 are sufficiently general to be adapted to a variety of decision contexts. In any event, it is important that the building design team arrives at a common understanding of how risk is to be measured since this will be required in evaluating whether the performance objectives for the project are met.

Investment in risk reduction invariably must be balanced against available resources. There currently are many such trade-offs in the area of fire protection (e.g., reductions in hourly ratings if sprinklers are provided), but the effect of the trade-off on performance is not quantified (Beyler 2004). Considering each of the hazards term by term in Eq. 3.1 (or hazard scenarios in Eq. 3.2), one can estimate the probability of unacceptable loss due to each hazard (or scenario) and the relative contribution of each to overall building risk. This overall risk must be limited to a socially acceptable value through a combination of professional practice and building regulation. Strategies for reducing risks from various hazards or fire scenarios may be directed toward different aspects of structural behavior and performance. For example, in earthquake engineering, such strategies would be aimed at the lateral force-resisting system, while for fire they would be aimed at enhancing the integrity of the structure subjected to gravity loads in a degraded or damaged condition. It is unwise to invest large sums in marginally reducing the risk from one hazard or scenario while others go unaddressed. Accordingly, the building design team must attempt to identify and document major sources of risk as part of the PBFE process. Those sources that contribute only trivially to risk should be screened out so that structural design and fire protection can focus on risk mitigation strategies that maximize the return (risk reduction) on investment. While the scenario analysis (Eq. 3.2) cannot provide a perspective on the overall risk to the building from fire in comparison to risk from other natural or man-made hazards, specific scenario risks are more easily evaluated and communicated among members of the design team and project stakeholders and their relative importance established. Furthermore, scenarios are far more easily implemented in building regulation. Accordingly, initial implementation of PBFE in structural engineering practice is likely to be scenario-based.

Efforts should be made to identify and analyze all uncertainties that affect the risk metric and to display them clearly in the risk assessment. Specific sources of uncertainty in analyzing structural response to fires would include:

- Scenario identification
- Fire load density
- Compartment ventilation
- Structural modeling
- Thermal and mechanical properties of steel and concrete
- Limitations in supporting databases

38

Uncertainties in engineering risk analysis come in two basic types, designated as aleatoric and epistemic. Aleatoric uncertainties represent inherent randomness that is irreducible (at the customary scales of engineering analysis) and cannot be eliminated by further analysis or testing. Fire load density, occupancy live loads, and strengths of structural frames are examples of engineering parameters that are inherently random. Such parameters are represented by probability distributions, and the uncertainties are propagated through the risk analysis (Eqs. 3.1 and 3.2) to yield a point estimate of risk, i.e., "The probability of building collapse is less than 10^{-5}/yr." In contrast, epistemic or knowledge-based uncertainties depend on the engineering models and supporting databases and can be reduced (at additional cost) by using improved or advanced (and usually more complex) models and more complete databases. When these uncertainties are propagated through the risk analysis, they yield an "interval" estimate of risk (in a Bayesian sense, since the databases invariably are limited): "I am 90 % confident that the probability of building collapse is less than 10^{-5}/yr." While a statement of such probabilities may not be required as part of every design decision, displaying the epistemic uncertainty in the risk assessment is important. The display is a manifestation of the confidence in the modeling that supports the decision process. Either overstating or understating the epistemic uncertainty can distort the risk mitigation policy.

3.3 FIRE HAZARD MODELING

Section 3.2 provided a general framework for assessing the risk to building construction from natural and man-made hazards, including fires. In this section, these methods are considered as they apply specifically to the assessment of fire risk. According to the National Fire Protection Association, there were approximately 1.7 million fire starts reported in the United States in 2002, leading to approximately 3,400 fatalities and property damage of $10 billion. These risks were not randomly distributed across the building inventory, however, since approximately 80 % of the fatalities and 75 % of the economic losses were concentrated in the residential sector. Single-family residences and low-rise multifamily apartments typically are not engineered and thus would fall outside the scope of the risk analysis and engineering solutions proposed in these guidelines. These guidelines are directed toward engineered construction in residential/commercial buildings, offices, public assembly facilities, schools, hospitals, and similar occupancies.

The fire hazard, defined by the terms P[H] or the postulated scenarios, H_s in Eqs. 3.1 and 3.2, depends on the incidence of fires, as well as the fire load (through the building occupancy), compartment dimensions, thermal and ventilation characteristics, and operability of active fire suppression, smoke-control, and air handling systems (sprinklers and smoke and heat vents).

3.3.1 Incidence of Fires

The random occurrence of rare (accidental) events such as fires is commonly modeled as a Poisson process. In its simplest formulation, if the occurrences of events are statistically independent, the probability of occurrence in any interval of time, Δt, is $\lambda \Delta t$, and the probability of two or more simultaneous events is essentially zero, then if $N(t) =$

the number of events that occur in (0,t), the probability that r events occur in that interval is (Stewart and Melchers 1997)

$$P[N(t) = r] = (\lambda t)^r \exp(-\lambda t)/r!; \quad r = 0,1,2,3\ldots \tag{3.3}$$

It may be shown that the expected (average) number of events in (0,t) is λt; hence λ = mean rate of occurrence of the Poisson events. Furthermore, the probability of at least one event in (0,t) is

$$P[N(t) > 1] = 1 - P[N(t) = 0] = 1 - \exp(-\lambda t) \tag{3.4}$$

If the events are very rare, then the probability of one occurrence in (0,t) is approximately equal to λt. Accordingly, if the probabilities and frequencies of occurrence of the hazard are annualized, $P[\text{Fire}] \approx \lambda_{\text{Fire}}$. As noted previously, it is questionable to model deliberate acts that target specific buildings for sociopolitical impact by this approach since such acts do not occur randomly in the building population at large.

The ignition of fire can be modeled as a Poisson event, with a mean rate of occurrence, λ_{Ign} that is related to floor area, A_f. This mean rate of occurrence is summarized in Table 3.2 for several common building occupancies (CIB W14 1983; SFPE/SEI 2003). The values presented were selected for illustration of what might be typical for a broad occupancy category and may not be accurate for specific buildings. The mean rate is on the order of 0.5 x 10^{-6}/m^2/yr to 1.0 x 10^{-6}/m^2/yr for common occupancies. There is evidence that the ignition rate is proportional to $\sqrt{A_f}$ rather than A_f in very large compartments, where the fuel load may not be uniformly distributed over the floor area. The rate of ignition may also depend on the age of the building and level of maintenance (Bennetts and Thomas 2002), which are not reflected in the table. Following ignition, the likelihood of a fully developed compartment fire with the potential to cause significant structural damage depends on the presence and timely activation of fire and smoke detection and suppression systems and quick response of the fire department. For example, if the reliability of the sprinkler system is on the order of 0.9 – 0.95 (Beyler 2004) and fire department response times are typical of urban areas, the probability of a fully developed fire in urban hotels and schools, given ignition, is on the order of 0.01 (CIB W14 1983). Thus, in modern code-compliant building systems, such conditions occur in a relatively small percentage of cases. The occurrence of full-development fires is also described by a Poisson process (with random selection), with mean rate of occurrence equal to the product of P[Full development|Ignition] and λ_{Ign}. This rate, denoted λ_H, is on the order of 10^{-8}/m^2/yr and is equivalent to the term P[H] in Eq. 3.1.

TABLE 3.2. Fire Occurrence

Occupancy	Mean rate x 10^{-6}/m^2/yr	P[Flashover|ignition]
Office	1 to 2	10^{-3} to 10^{-2}
Dwelling	2 to 5	10^{-1}
Hotel	0.5	10^{-3} to 10^{-2}
Commercial	1.0	10^{-2}
School	0.5	10^{-3}

While this annual frequency is order of magnitude only, it is not inconsistent with fire occurrence statistics reported independently. For example, Clifton and Feeney (2004) remark that the probability of a fully developed fire occurring over the 50-year life of a multi-story office building with sprinklers in New Zealand is on the order of 0.5 %. For a multi-story building with 10,000 m² of leasable space, this corresponds to an annual frequency of approximately 1×10^{-4}/yr, about what one would expect from the above formulation. Similarly, Bennetts and Thomas (2002) report a "fire start frequency rate" of approximately 2×10^{-3}/apartment/year in the United States. At an average apartment area of 110 m², this corresponds to an annual frequency of 1.8×10^{-5}/m²/yr, an incidence that is one order of magnitude higher than that in Table 3.2 for "dwellings," which includes both single-family residences and apartments. It often is sufficient in risk analysis to estimate such frequencies to an order of magnitude to evaluate engineering alternatives.

3.3.2 Characteristics of Fire Exposure Curves

Once a fire has ignited, its subsequent severity and impact on the structural system depend on the fire load (amount, type, distribution, and surface characteristics), compartment ventilation, compartment geometry, and thermal characteristics of the compartment bounding surface. The ASTM E119/E1529/ISO exposures describe a long-duration moderately severe post-flashover fire but do not address specifically any of the factors above. The monotonic increase implies that the supply of fuel is inexhaustible and there is no cooling phase. For example, in ISO Standard 834, the time-temperature curves are given by

$$T(t) = 20 + 345 \log_{10}(1 + 8t) \qquad \text{(cellulose)} \qquad (3.5a)$$

$$T(t) = 20 + 1080 \left[1 - 0.325 \, e^{-0.167t} + 0.675 \, e^{-2.5t} \right] \quad \text{(hydrocarbon)} \qquad (3.5b)$$

in which T is given in °C and t is in minutes. It should be noted that a temperature-controlled standard test such as that specified by ASTM (2002)or ISO may not imply a constant compartment fuel load since structural elements with high thermal mass require more fuel to maintain the furnace temperatures than lightweight construction materials. Furthermore, combustible structural assemblies contribute to the fuel load, reducing the fuel otherwise required to maintain furnace temperatures, making their comparison with the performance of non-combustible assemblies difficult.

The primary factors affecting fire development and the temperature history in a building compartment during a real fire are the fire load per unit area, $q_t = WM/A_t$; the ventilation parameter, $A_o \sqrt{h}$; the opening factor, $F = A_o \dfrac{\sqrt{h}}{A_t}$, which controls the rate of combustion; and the thermal properties of the compartment lining materials (CIB W14 1986; SFPE 2004). Parameter A_o denotes the total area (m²) of door and window openings in the compartment, h = average height (m) of the openings, A_t is total area (m²) of the compartment bounding surface, M is total mass (kg) of combustibles in the compartment,

41

and W = effective heat of combustion, which for a building fire involving normal building contents is taken as 18.6 MJ/kg (8,000 Btu/lb). (Chemical fires may have a different W, depending on the nature of the fuel.) When a building contains materials that pose a high fire hazard, involving high fuel loads, flammable liquids, or explosive materials, a special analysis of fire hazard and fire exposure (temperature and duration) should be conducted.

Fire load surveys have provided data to describe fire loads for common occupancies, often in terms of fire load density obtained from the calorific value of combustible material. These data usually are reported in terms of fire load per unit floor area, q_f, although occasionally they are reported as load, q_t, with respect to total area of the compartment bounding surface. The fire load is random in intensity and spatial distribution, but generally is reported as an equivalent uniformly distributed load with respect to floor area. A histogram for fire load in a general/clerical office in the United States (equivalent weight of combustibles having a calorific value of 18.6 MJ/kg (8,000 Btu/lb), (Culver 1976) is shown in Figure 3.1. Tables 3.3(a) through 3.3(c) summarize some of the fire load data from the United States and Europe that might be considered in selecting a fire load, q_f (MJ/m^2), with respect to compartment floor area. Appendix 1 of (CIB W14 1986) summarizes fire loads, q_f, from a variety of studies in Europe and the United States. There are significant variations in q_f from occupancy to occupancy and from study to study for a given occupancy. For example, for general offices, the mean value of q_f (in MJ/m^2) = 420 (ECCS 2001), 390 (CIB W14 1983), 598 (Culver 1976), and 348 (Kumar and Rao 1997). As a further benchmark for comparison, the Cardington fire tests (Newman, Robinson, and Bailey 2000) were conducted with fire loads of 372 MJ/m^2 to 744 MJ/m^2 (floor area), assuming an effective heat of fuel of 18.6 MJ/kg. The data summarized in Table 3.3 present the mean and standard deviation of q_f and its 80th and 90th percentiles. (The ECCS 111 (2001) has recommended that the design fire load be taken at the 80th percentile.) It is notable that the coefficient of variation (COV) in fire load is 0.30 for all occupancies in the ECCS 111 recommendations, a value that is substantially less than that measured in both Culver's (1976)

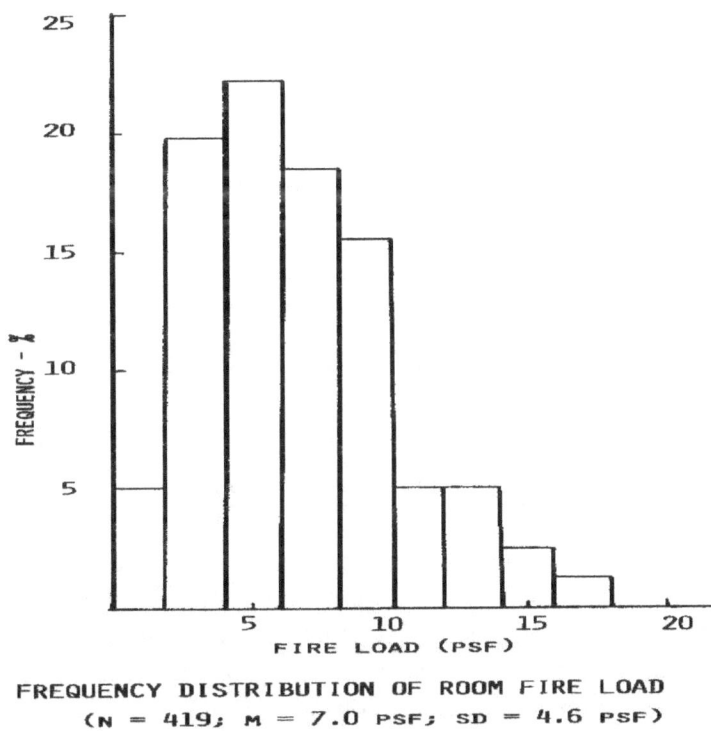

FREQUENCY DISTRIBUTION OF ROOM FIRE LOAD
(N = 419; M = 7.0 PSF; SD = 4.6 PSF)

FIGURE 3.1. Underrated Fire Load with Respect to Floor Area in General/Clerical Offices in the United States (Culver 1976) [1 psf = 47.88 Pa]

and Kumar and Rao's (1997) surveys (approximately 0.60). Moreover, it is substantially less than the COVs in load intensities measured in most modern surveys of live loads in the United States and Western Europe, which tend to be on the order of 0.50 to 0.60, depending on the loaded area.

TABLE 3.3(a). Fire Load with Respect to Floor Area (MJ/m^2) (CIB W14, 1983; 1986)

Occupancy	Mean	Standard deviation	80%ile	90%ile
Offices	420	309	680	740
Dwellings (bedroom)	640	135	750	810
Hotels	345	92	420	472
Schools (primary)	285	79	360	415

TABLE 3.3(b). Fire Load in Offices with Respect to Floor Area (MJ/m^2) (Culver 1976)

Occupancy	Mean	Standard deviation	80%ile	90%ile
General/clerical offices	598	358	898	1,046
Conference rooms	425	425	714	969
File, storage rooms	1,112	1,020	1,968	2,400

TABLE 3.3(c). Fire Load with Respect to Floor Area (MJ/m^2) (ECCS 2001)

Occupancy	Mean	Standard deviation	80%ile	90%ile
Office	420	126	511	584
Dwelling	780	234	948	1,085
Hotel	310	93	377	431
Shopping center	600	180	730	835
Schools	285	86	347	397
Hospitals	230	69	280	320
Theaters	300	90	365	420

The fire loads reported above are obtained, for the most part, from building surveys. By their nature, such surveys capture the fuel load at an arbitrary point in time. It should also be noted that fuel loads, like other occupancy loads, vary during the life of a building as tenancies and occupancy classifications change. Fuel loads during periods of remodeling, rehabilitation, or construction may be relatively high, at a time when protection systems may not be functioning. Such loads would not be reflected in Tables 3.3(a) through 3.3(c). Clearly, there is a need for further research on fire loads for common building occupancies.

To evaluate the time–temperature curve for a fully developed (post-flashover) fire resulting from a given fire scenario, a post-flashover condition can be assumed in which thermal conditions are homogeneous throughout the compartment, the fire is ventilation controlled, and no combustion occurs outside the compartment. Most post-flashover fires satisfy these conditions. The volumetric flow of air into the compartment is proportional to the ventilation factor $A_o \sqrt{h}$, in which A_o = total area (m^2) of door or window openings in the compartment and h = average height of openings; thus, the total rate of heat release within the compartment is also proportional to $A_o \sqrt{h}$. The duration of the burning phase of the fire then is proportional to $MW/(A_o \sqrt{h})$. A solution to the energy balance equation then yields the compartment temperature, T, as a function of time, t, in terms of opening factor F, q_t, and the thermal properties of the compartment interior surface, which defines the fire exposure (e.g., Magnusson and Thelandersson 1974; SFPE 2004). Such relations have been verified experimentally. To illustrate, it was assumed that $A_t = 4A_f$, noting that for typical rooms the total surface area is three to five times the floor area. Based on these data, two fire exposure curves were developed to simulate a compartment fire test conducted at the National Bureau of Standards (now National Institute of Standards and Technology). The fire load was q_t = 100 MJ/m^2 ($q_f \approx$ 400 MJ/m^2) and the two opening factors believed to be typical in light occupancies: F = 0.04 m$^{1/2}$ and 0.08 m$^{1/2}$. These fire exposures are compared to the standard ASTM E119 (2007) exposure and the temperatures measured in the fire test in Figure 3.2. Temperatures measured in the compartment fire increase more rapidly after flashover than indicated by the ASTM standard fire exposure, but the fire duration is relatively short and the temperatures decay for an extended period (fuel-controlled phase). Knowledge of the fuel load and compartment ventilation can be used to generate a fire exposure that matches a natural fire quite well. Such fires can be approximately parameterized (Buchanan 2001; ECCS 2001; SFPE 2004).

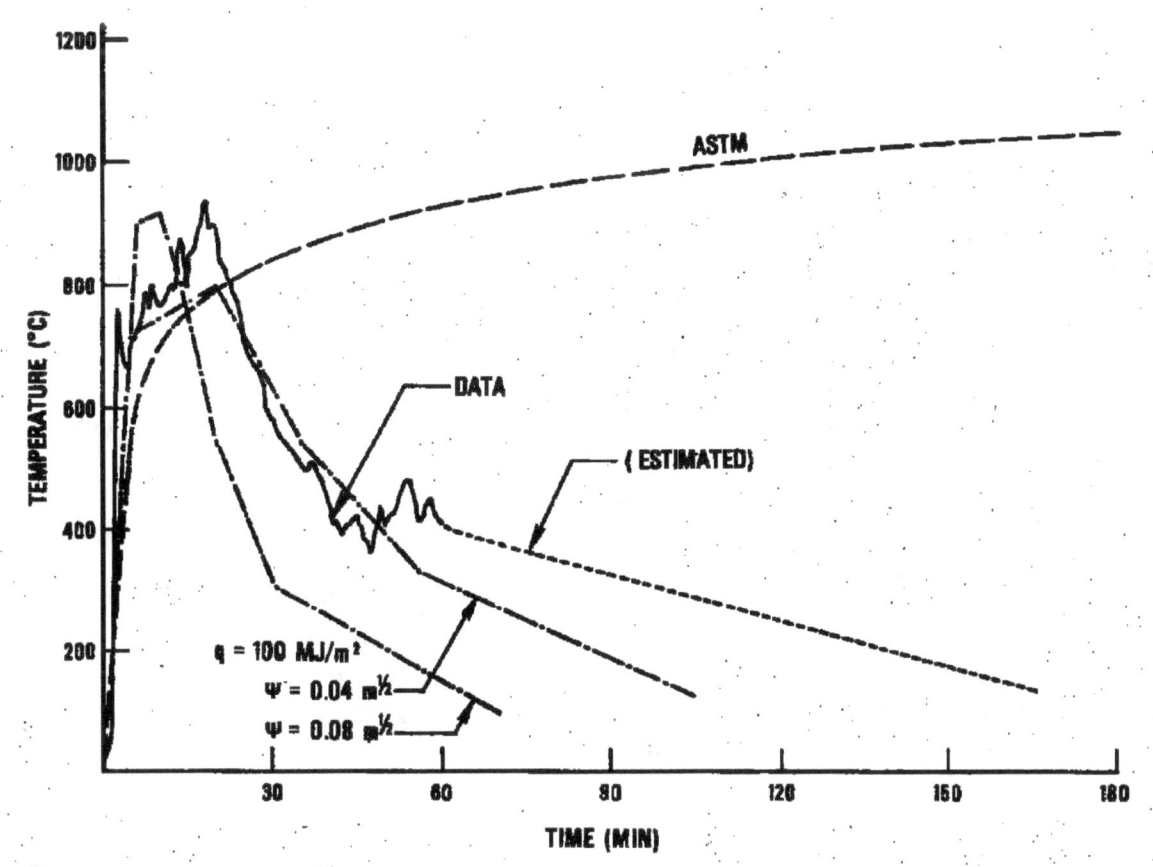

FIGURE 3.2. Comparison of Calculated Fire Exposures to ASTM E119 and Compartment Fire Test. (Fire load is referenced to total compartment bounding area, A_t.)

The above analysis yields one fire exposure curve for a compartment. The fuel source, arrangement, and load, and the compartment ventilation and composition of its bounding surfaces constitute one fire scenario. A "scenario" is a set of conditions—fire protection, ignition, nature/configuration of fuel, ventilation, etc.—that define the fire exposure curve used in design (SFPE 2000). In structural engineering of frames for fire protection in most multi-tenant buildings, several fire scenarios may have to be considered. This is considered further in Section 3.4.3.1.

3.4 STRUCTURAL ENGINEERING FOR FIRE CONDITIONS

This section considers the structural integrity aspects of fire protection. Once the design-basis events are determined, the principles of structural analysis used in designing for fire conditions are similar to those used for other hazards, but the conditional nature of the limit states leads to some differences that will become apparent in the following discussion. General design strategies require a statement of performance objectives and a general approach to risk management. Specific technical engineering approaches then can be developed along two lines: design by engineering analysis and design by qualification testing.

3.4.1 Performance Objective

The primary performance objective underlying structural engineering for fire conditions is that of life safety. Fire safety levels should depend on the nature of the fire hazard and building occupancy, height of the building, presence of active and passive fire mitigation measures, and the effectiveness of firefighting. Given the occurrence of a fire, three limit states traditionally have been considered (Buchanan 2001):

1. Heat transmission leading to unacceptable rise in temperature on unexposed surfaces
2. Breach of barrier due to loss of integrity
3. Loss of load-bearing capacity (or, for short, insulation/integrity/stability)

Other specific performance objectives must be determined by the stakeholders in the building process. Performance-based engineering, in contrast to traditional prescriptive approaches that essentially follow a design recipe, involves goal-setting and documentation on the part of the design team. Performance objectives for fire-resistant design might include, but are not necessarily limited to the following (e.g, ICC 2006; NFPA 2003; CIB W14 2001):

- Life safety of building occupants
- Life safety for firefighters entering the building
- Survival of burn-out of building contents
- Preservation of paths of entrance/egress
- Protection of property (including adjacent buildings) or minimal disruption of business operation
- Protection of the environment
- Protection of civil infrastructure

3.4.2 General Fire Risk Mitigation Strategies

Fire risk mitigation and fire safety measures can be aimed at three levels:

1. Preventing the outbreak of fires through elimination of ignition sources or hazardous practices

2. Preventing fire development (flames and smoke) through early detection and suppression
3. Preventing loss of life or structural collapse through provision of general structural integrity, compartmentation, fire protection systems, and other measures

Proper management of building risk involves examining each of the terms in Eqs. 3.1 or 3.2 as part of the decision process. Structural engineering for fire is concerned primarily with maintaining compartmentation (limiting spread) and general structural integrity.

3.4.2.1 Event Control/Fire Protection Systems

Event control focuses on the term P[H] in Eq. 3.1. One principle of risk management of man-made hazards is that it usually is more cost-effective to prevent or control a situation than to deal with one that has progressed beyond the initial stage. In other words, if P[H] $\approx \lambda_H$ can be limited to (approximately) the *de minimis* threshold, then further structural engineering measures for fire protection to reduce the other terms in the risk equation are not necessary. For many types of building construction this may be the most cost-effective strategy for risk mitigation. If one building performance objective is to protect the building and its contents from significant damage (for example, in a financial institution, record-keeping facility, or museum), or to minimize the likelihood that damage to adjacent buildings may occur (as in a historic district), it is important to minimize the likelihood of a major fire ever developing. High levels of structural integrity may not be particularly useful for such applications if the contents are essentially irreplaceable or costs of contents damage are significant.

Conversely, if P[H] is one or two orders above the *de minimis* threshold, further investigation of that hazard is warranted. (It is interesting to note that the National Building Code of Canada (1996) specifies a threshold for the consideration of "abnormal" loads at 10^{-4}/yr. If the likelihood of the event is greater than that threshold, consideration of the event must be documented.) The evaluation of P[H] has several dimensions: likelihood of ignition, flashover, and behavior of the building occupants when confronted with the threat. The effectiveness of nonstructural measures, such as sprinklers, vents, and intervention, first by occupants and subsequently by firefighters once they arrive on the scene, are taken into account in the risk analysis through the term P[H] in Eq. 3.1. Collapsing these factors into one term may appear to oversimplify what may be a complex systems analysis. An event tree formulation may be helpful in visualizing the alternatives and in identifying plausible scenarios (Mowrer 2004). Such an event tree should include the probability that the sprinklers malfunction, the probability that the building ventilation system is operating properly, and the likelihood that emergency personnel arrive at a state where the fire can still be controlled, and the probability that passive protection will not have been damaged or destroyed by ancillary events.

3.4.2.2 Structural Engineering Analysis

Structural engineering measures to design fire resistance into the building structural system directly affect the terms P[LS|H] and P[D|LS] in Eq. 3.1. For simplicity in structural code development (e.g., AISC 2005), the term P[Loss> θ|D] P[D|LS] in Eq. 3.1 may be collapsed into one conditional probability, P[DS|LS], in which DS = damage state identified as being consistent with the performance objectives for the building project. The damage state might be stipulated as "moderate damage" or "collapse," depending on the performance objective. If the overall risk (measured by annual frequency) is to be limited to less than 10^{-6}, then the engineering measures for fire protection must ensure that

$$\Sigma_{LS}P[DS|LS]P[LS|H] < 10^{-6}/\lambda_H \tag{3.6}$$

This criterion can be satisfied with a combination of passive fire protection and structural engineering measures, as described in the following section. Consistent with the idea of risk differentiation among different building occupancies that is implied in Table 3.1, one might specify that 10^{-6} in Eq. 3.6 be replaced by 10^{-5} for buildings in which the consequences are less severe and with 10^{-7} for essential facilities that cannot be evacuated, such as hospitals. Such risk differentiations are made only indirectly in current building regulations, if at all (through such devices as "importance factors"), and should be carefully debated before implementation.

3.4.3 Structural Design by Engineering Analysis

A structurally significant fire results in imposed deformations on the building structure, and the forces developed in an indeterminate building frame are self-limiting in nature. Moreover, the strength and stiffness of structural materials are temperature dependent (degrading to only a fraction of the ambient strength and stiffness at temperatures as high as 1,000°C) in a fully developed compartment fire. Results of recent full-scale building fire tests (Newman, Robinson, and Bailey 2000) indicate that advanced structural analysis can reproduce the actual behavior of the structure quite well.

3.4.3.1 Design-Basis Fire Scenarios

A scenario identifies a set of conditions—sources of ignition, nature and configuration of fuel, ventilation, patterns of growth and spread of smoke, availability of active fire detection and protection systems, etc.—that can be used as a basis for verifying that the performance objectives of the design have been achieved. In risk analysis and management, scenario analyses are commonly used to answer "what if" questions. It is not possible to identify all possible scenarios, but an effort must be made to identify those that are dominant contributors to fire risk.

The performance option in the NFPA 5000 *Building Construction and Safety Code* stipulates eight scenarios (Section 5.5.2), but not all must be evaluated fully (Section 5.5.1.3). Section 1701.3.15 of the *ICC Performance Code* requires scenarios that "can be reasonably expected to impact in buildings as designed or constructed" but does not require a specific number of scenarios. The number of design scenarios must be kept to a manageable size since the number of possible scenarios may become quite large in complex facilities. Stipulating an excessive number of scenarios (e.g., involving ignition and development in different parts of the building) may discourage engineers from applying PBFE and cause them to regress to traditional prescriptive or deemed-to-satisfy provisions.

Each fire scenario ultimately produces a fire exposure curve that can be used in advanced structural analysis to assess the structural response during the heating and cooling phases following exhaustion of combustibles. Typically, three types of fire scenario might be considered (Hurley 2004). The first and most important would involve a fully developed (post-flashover) fire with involvement of all combustibles within the compartment (similar to the one illustrated in Figure 3.2). This scenario would stipulate the amount and distribution of fuel, compartment ventilation, and the type of materials forming the floor, walls, and ceiling of the compartment. The second, and related, scenario involves window flames that escape the enclosure of a post-flashover fire to heat exterior structural elements. The third would consider the impact of local fire plumes adjacent to exposed and unprotected structural elements; this scenario may occur in large compartments where fuel is not distributed uniformly and there is little interaction with the enclosure. The SFPE *Engineering Guide: Fire Exposures to Structural Elements* (SFPE 2004) provides additional information on the analysis of local fire plumes. Scenarios should include the possibility that sprinklers (if present) fail to function. At a greater level of sophistication, some scenarios may involve fires spreading to several adjacent compartments or floors in the course of fire development. The development of fire exposure curves for multiple-compartment fires is highly complex (Beyler 2004). In the last decade, computational fluid dynamics (CFD) models have become widely used by fire protection engineers to simulate the behavior of fires in buildings (SFPE Handbook, 4th ed.). While there is extensive validation of these models in situations where the fire's heat release rate is known or specified (NUREG-1824), there has been less validation work for post-flashover fires, fire growth and spread, and under-ventilated fires.

3.4.3.2 Thermal Effects on Structural Components and Systems

The fire exposure, nature of heat transfer, and thermal properties of the material govern the development of temperatures (and thus thermal strains) in the structural components and system. If the temperature is essentially uniform in the compartment, these temperatures can be determined from the numerical solution of the heat flow (diffusion) equation. Assuming heat flow in two dimensions and homogeneous and isotropic thermal conductivity,

$$\rho C_p \, \partial T / \partial t = k \, \nabla^2 T \tag{3.7}$$

in which $T = T(x,y,t)$ represents the temperature within the structural element at coordinate (x,y), ρ = density, thermal parameters C_p = specific heat (typically 600 J/kg K for steel and 900 J/kg K for concrete), and k = thermal conductivity (typically 45 W/m K for steel and 1.4 W/m K for concrete). The thermal parameters may be temperature dependent (ECCS 2001). Assuming that the heat flow across the boundary due to the fire is caused by both convection and radiation, the boundary condition at the exposed surface for the solution of Eq. 3.7 is

$$-k \, \partial T / \partial n \; = \; h_c (T_g - T_s) \; + \; V \varepsilon \sigma \, (T_g^4 - T_s^4) \tag{3.8}$$

Where:

n	=	Direction of heat flow at the boundary
h_c	=	Convection coefficient (typically 25 $W/m^2 K$)
V	=	Radiation view factor
ε	=	Resultant emissivity (typically about 0.8 for hot surfaces and luminous flames)
σ	=	Stefan-Boltzman constant (5.67 x 10^{-8} $W/m^2 K^4$)
T_g, T_s	=	Absolute temperatures of compartment (e.g., Figure 3.2) and surface, respectively

Analytical solutions to Eqs. 3.7 and 3.8 are difficult to obtain, and finite element or finite difference approaches generally are necessary for all but simple structural member tests. In the finite element formulation, Eqs. 3.7 and 3.8 become,

$$\mathbf{C} \, \partial \underline{T} / \partial t + \mathbf{K} \, \underline{T} = \underline{Q} \tag{3.9}$$

Where:

$\mathbf{C}$	=	Heat capacity matrix
$\mathbf{K}$	=	Thermal conductivity matrix
$\underline{Q}$	=	External heat input
$\underline{T}$	=	Temperature vector

A number of programs are available to perform the thermal analysis (Milke 2002).

Simplified analytical and graphical methods are available for computing temperature develop-ment in simple beams and columns. Some of these methods are summarized in T.T. Lie et al. (1992).

3.4.3.3 Strength Requirements and Deformation Limits

Limit states design is well accepted for structural engineering for occupancy and environmental loads. Structural engineering for fire conditions must adopt a similar approach. The structural actions resulting from the design-basis fire or alternative fire scenarios must be integrated into the structural analysis and design process. This section summarizes the probabilistic basis for appropriate combinations of loads to facilitate fire-resistant structural design and recommends specific load combinations for this purpose (Ellingwood and Corotis 1991; Ellingwood 2005a). The probabilistic basis is essential for measuring compliance with performance objectives, for comparing alternatives, and for making the role of uncertainty in the decision process transparent.

Structural loads vary randomly in space and time. Modern structural reliability theory has illuminated the analysis of load combinations, and has led to design load combinations that have a specified probability of being exceeded (ASCE 2005). The bases for the probabilistic modeling of loads and load combinations have been published in the archival literature (e.g., Ellingwood et al. 1982). To summarize, both theoretical analysis and simulation have shown that when two or more loads that vary in time are combined, the maximum combined effect, U, during time interval (0,t) occurs when one load achieves its maximum (or "principal" value during 0,t while the other loads are at their companion values. In other words, if one were combining the structural action due to the fire, T, with dead load, D, and occupancy live load, L, one would consider

$$U = \text{Max}\,[D + L + T] = D + \max\,[\max L + T, L + \max T] \qquad (3.10)$$

Equation 3.10 is a simple statement of the "principal action–companion action" load combination scheme that is the basis for all modern limit states structural codes worldwide (ASCE 7-05; *Eurocode 1* 1994). Since fire is a very rare event with a duration that measures in hours in any local region within the building, the probability that a fire occurs concurrent with the maximum live load is negligible—in fact, it is on the order of 10^{-9}/yr (Ellingwood and Corotis 1991; Ellingwood 2005 —so the term "max L + T" in Eq. 3.10 is simply equal to max L, which is addressed by the ordinary gravity load design criteria in ASCE Standard 7-05. The L in "L + max T" is the "companion action live load, or the value expected to act at the time of the design-basis fire, represented by max T. Surveys of live loads (e.g., Culver 1976; Ellingwood and Corotis 1991) show that the mean value of this companion action live load generally is on the order of 25 % to 30 % of the nominal live load specified in ASCE Standard 7-05.

Reasoning along these lines, one can develop a suitable set of load combinations involving fire and gravity load effects. Such combinations are required for limit states design based on advanced structural analysis but could also be used for designing a suitable fire test of a structural component or assembly. Observing the reliability constraint in Eq. 3.6 limiting the collapse probability to less than 10^{-6}/yr and noting that λ_H typically is on the order 10^{-5} to 10^{-4}/yr, the load factors in the principal action/companion action load combination are adjusted until the probability that the combined structural action, U, exceeds the design action, U_d, is equal to approximately 0.05 to 0.1. This iterative process yields

$$U_d = (0.9 \text{ or } 1.2)D + T + 0.5L + (0.5L_r \text{ or } 0.2S) \qquad (3.11)$$

in which T (replacing term "max T" for simplicity) denotes the structural action (force or deformation) resulting from the postulated fire scenario (discussed below). The loads D, L, L_r, and S are the nominal loads in ASCE Standard 7-05. As might be observed from the brief review of probabilistic load models presented above, the companion actions 0.5L, $0.5L_r$, and 0.2S represent the most probable values of load on the structure at the time of the fire. When gravity effects stabilize the structure, the load factor on D is 0.9 rather than 1.2, and the live load is set equal to zero. Equation 3.11 is found in Commentary 2.5 of ASCE Standard 7-05 (ASCE 2005) and in Appendix 4 of the new AISC 2005 Specification (AISC 2005). A similar relationship is found in *Eurocode No. 1* (1998) and in the *ECCS model code on fire engineering* (ECCS 2001). Probabilistic event combination analyses show that neither wind nor earthquake effects need to be considered in checking the overall behavior of a building frame during a severe fire (Ellingwood 2005a). It should be emphasized that because of the nonlinear nature of the structural system response to severe fires, the loads and structural actions should be factored *prior* to performing the structural analysis.

The structural action, T, is determined from an analysis of the fire exposures that arise from postulated fire scenarios for design. Since these are "design-basis" events, the load factor is set equal to 1.0. In the Eurocode and ECCS, it is recommended that T be developed from the 80th percentiles of the fire load densities, such as those presented in Tables 3.3(a) through (c); these fire densities are permitted to be reduced by as much as 50%, depending on the presence of active fire protection systems in the building. However, it should be noted that the role of various active protection measures in reducing fire risk is already taken into account, in an average sense, in the value of λ_H in Eq. 3.6 on which load combination Eq. 3.11 is based. In these circumstances, any adjustments to these fire densities should be viewed with caution.

The lateral stability of building frames normally is ensured by designing the frames for lateral forces from wind and/or earthquake effects. However, few building frames are perfectly symmetric or symmetrically loaded by gravity loads. Moreover, columns and beams are not perfectly straight, nor are fabrication and erection procedures perfect. Consequently, even a "perfect" frame is subject to sway. If this sway is not accounted for, or if the imposed deformations from the fire give rise to significant frame deformations, large secondary (P-Δ) forces will develop in the frame and lead to overall instability of the frame under gravity loads. This occurrence can be mitigated by stipulating that the lateral stability of the building frame be checked by imposing a small notional lateral force equal in magnitude to 0.002ΣP at each floor level, in which the term ΣP is the cumulative gravity force due to the summation of dead and live loads acting on the story above that level. This approach to ensuring lateral stability under gravity loads has been recommended by the Structural Stability Research Council (SSRC 1998) and is being implemented in several modern standards (AISC 2005; NBCC 2005). Structural elements that provide lateral stability and are exposed to heating from the fire should receive particular attention in fire-resistant structural design.

The above load combinations determine the required strength of the building frame (strength that must be provided in design) from structural analysis. The design strength is determined by (ACI 2005; AISC 2005):

$$\text{Required strength } (U_d) < \text{Design strength } (\phi R_n) \tag{3.12}$$

in which R_n = nominal strength stipulated in the material specification or code (e.g., strength in tension, flexure, shear, or compression) and ϕ = resistance factor that takes into account uncertainties in the determination of R_n and mode and consequences of failure. The design strength and deformations should be calculated taking the elevated temperature properties of the structural materials into account (AISC 2005; ACI 2005). The stability check of the frame should include second-order forces arising from differential heating of the structural system. The selection of specific resistance factors for governing structural limit states is the responsibility of standard-writing groups for the individual construction materials and is outside the scope of this chapter. Although some have suggested that the resistance factors for fire (and other accidental loads) should be set equal to 1.0 (ECCS 2001), the structural properties of steel and concrete at the elevated temperatures in a severe fire are more uncertain than the corresponding properties under normal conditions. Engineers who would err in the direction of conservatism might use resistance factors that are the same as those used for normal design.

3.4.3.4 Required Attributes of Advanced Analysis

The analysis must include both thermal response and mechanical response of structural components and systems. Thermal and mechanical properties of structural materials are temperature-dependent (Gustaferro and Martin 1989; ECCS 2001), as described in detail in Chapters 4 for steel and 5 for concrete. Unfortunately, there currently is no standard ASTM test for determining structural properties at fire temperatures. The deterioration in structural strength and stiffness with increasing temperatures, nonlinear material behavior, effects of thermal expansion, and large deformations should be taken into account. The appropriate limit states include excessive deflections, connection fractures, and overall and local buckling. The analysis should allow for the nature of the failure observed in fire tests of structural systems (Lim et al. 2004). For example, the Cardington tests showed that, in structural frames with fire-protected columns and floor slabs supported by unprotected steel beams, the floor systems supported the load through the development of two-way membrane action rather than flexural action. Thus, the analysis should take such behavior into account. Furthermore, for fire-induced limit states that are relevant for progressive collapse mitigation, a finite element (FE) platform with nonlinear analysis capabilities generally would be required for structural systems analysis.

3.4.3.5 Simplified Methods of Analysis

Simplified methods of analysis may be sufficient when structural design for fire conditions involves structural members and components rather than systems and when the element is exposed to essentially uniform heat flux on all sides (e.g., Milke 1985; Lie and Almand 1990; Lie 1992; Buchanan 2001; Bennett and Thomas 2002; Bailey 2004). Some general guidelines are provided in Appendix 4 of the AISC *Specification* (2005), in Chapter X of *AISC Steel Design Guide* 19 (Ruddy et al.2003), and in the recommendations of ACI Committee 216 (2007). ASCE/SFPE Standard 29-05 (ASCE 2005) also contains a number of relatively simple analytical methods. However, Section

1.2.1 of *ASCE/SFPE Standard 29-05* states that the methods are intended to produce fire resistance rating times for concrete, timber and wood, masonry, and steel construction that are equivalent to those obtained from a standard ASTM E119 fire test, and to produce results that are an alternative to laboratory test results. ASCE/SFPE Standard 29-05 also states (Section 1.3.1) that it may provide results that do not describe the performance for natural fires. Accordingly, the methods in ASCE/SFPE Standard 29-05 may not always be suitable for demonstrating compliance under natural fires.

3.4.4 Structural Design by Performance-Based Testing

Design by testing is an acceptable alternative to design by analysis in performance-based fire engineering. Fire testing in support of PBFE should demonstrate compliance with the performance objectives, a different goal than that in traditional fire testing which is focused on demonstrating that the component tested meets the three criteria in ASTM Standard E119 during the required rating period. In contrast, performance-based testing should strive to replicate the same basic attributes—fire exposure, superposed gravity loads, conditions of restraint—identified as being essential in advanced analysis. Such tests of building structural systems seldom would be feasible as a design tool, but tests of key structural components could be considered and performed on a case-by-case basis. While procedures for performing such a test have yet to be standardized, the move toward an acceptance of performance-based fire engineering by structural engineers and the code community may provide incentives for their development.

3.5 SUMMARY

Practical design solutions can be developed to achieve performance-based engineering objectives for fire resistance, and protection can be expressed in terms of acceptable risk. Adoption of PBFE as an alternative approach to fire safety assurance would have a number of benefits. Foremost among these is the flexibility it provides in designing to meet mutually agreed-upon building performance objectives including, but not necessarily limited to, life safety and property protection for specific building occupancy categories and anticipated risks. The ability to consider fire protection alternatives, to trade off investments in additional fire protection above the code minimums against benefit received, and to reduce or eliminate unnecessary fire protection would add significant economic value to design of certain building structural systems.

Performance-based fire engineering requires a different approach to building planning and design from what is customary as well as better integration of the stakeholders, building design team, and the building code community. At the outset, a thoughtful approach must be taken to establishing performance objectives and identifying possible fire scenarios to check compliance with those objectives. This requires a systems approach, and involvement of project stakeholders early in the planning stages of design, and a degree of integration of traditional architectural and structural design functions. In the technical arena, the structural engineering task will become more complex. Advanced analysis methods may be warranted for major buildings or where additional investment in design is warranted by the nature of the building project or can be shown to have a major economic benefit. Large-scale fire testing has demonstrated the power of modern advanced analysis methods. Many of the analytical (thermo-structural modeling) tools are becoming accessible, and a more widespread demand for such tools will make them more user friendly. On the other hand, not all PBFE solutions require advanced analysis; some can be achieved using relatively simple structural calculations. Finally, PBFE should include an assessment of uncertainties and must carry with it recognition of the need to make trade-offs between performance and cost.

Improving building fire safety by adopting PBFE methods is likely to provide economic incentive. For the majority of buildings, current methods of fire protection and demonstration of code compliance appear satisfactory and will continue to be used. On the other hand, PBFE gives the building design team and structural engineer additional quantitative tools for fire safety assurance in situations where prescriptive limits found in traditional codes may be highly restrictive or unsuitable, where safety benefits may be realized for unique facilities by better quantification of their structural fire resistance, or where innovative architectural expressions can be inhibited by customary fire resistance rating requirements. In addition, nonconforming fire code issues with existing construction can be addressed efficiently with PBFE prior to undertaking costly rehabilitation. To take full advantage of these new tools, structural engineers will have to develop a competency for fire-resistant structural analysis and design through education, and become comfortable in accepting this additional design challenge and responsibility.

3.6 REFERENCES

Fire risk assessment of buildings is an interdisciplinary endeavor. The past two decades have seen the development of an extensive literature on risk assessment and the performance of building structural systems exposed to severe fires. Structural analysis and decision tools and experimental data on performance of structural components and systems have been covered extensively, and the field continues to evolve rapidly. The following bibliography, tailored to a structural engineering audience, provides a point of departure for further study of the field. Many of the references listed are cited in this chapter. References that are suitable as a starting point for independent study are indicated with an asterisk.

ACI (2005), "Building Code Requirements for Structural Concrete," ACI Standard 318-05, Farmingham Hills, Mich.: American Concrete Institute.

ACI Committee 216 (2007), "Code Requirements for Determining Fire Resistance of Concrete and Masonry Construction Assemblies," ACI 216.1-07/TMS 0216.1-07, Farmington Hills, Mich.: American Concrete Institute.

AISC (2005), *Specification for Structural Steel Buildings*, Chicago American Institute of Steel Construction, Inc.

ASCE (2005), *Standard Calculation Methods for Structural Fire Protection*, ASCE/SFPE Standard 29-05, Reston, Va.: American Society of Civil Engineers.

ASCE (2005), *Minimum Design Loads for Buildings and Other Structures*, ASCE Standard 7-05, Reston, Va.: American Society of Civil Engineers.

ASTM (2007), *Standard Methods of Fire Tests of Building Construction and Materials*, ASTM Standard E119-07a, West Conshohocken, Pa.: American Society for Testing and Materials.

Bailey, C.G. (2004), "Membrane Action of Slab/Beam Composite Floor Systems in Fire," *Engrg. Struct.* 26(12): 1691-1703.

*Bennetts, I.D., and I.R. Thomas, "Design of Steel Structures Under Fire Conditions," *Progress in Struct. Engrg. and Materials* 4:1 (2002) 6-17.

Beyler, C. (2004), "Relationship Between Structural Fire Protection Design and Other Elements of Fire Safety Design," *Proc. NIST-SFPE Workshop for Development of a National R & D Roadmap for Fire Safety Design and Retrofit of Structures*, NISTIR 7133:100-106, April.

*Buchanan, A.H. (2001), *Structural design for fire safety*, Chichester, U.K.: John Wiley and Sons, Ltd.

CIB W14 (1983), "A conceptual approach towards a probability-based design guide on structural fire safety," *Fire Safety Journal* 6:1 (1983) 1-79.

CIB W14 (1986), "Design guide—structural fire safety," P.H. Thomas, Coordinator, *Fire Safety Journal* 10:2 (1986) 75-137.

*CIB W14 (2001), "Rational safety engineering approach to fire resistance of buildings," CIB Report No. 269, Rotterdam: Int. Council for Research and Innovation in Building and Construction, 2001.

Clifton, G.D., and M.J. Feeney (2004), "Fire Engineering Applications to Multi-Story Steel Structures," *Modern Steel Construction*, Chicago: AISC, March 2004.

Culver, C.G. (1976), "Survey Results for Fire Loads and Live Loads in Office Buildings," *Building Science Series No. 85*, Washington: National Bureau of Standards.

*ECCS Technical Committee 3 (2001), *Model code on fire engineering*, Document No. 111, Brussels: European Convention for Constructional Steelwork.

Ellingwood, B.R., et al. (1982), "Probability-Based Load Criteria: Load Factors and Load Combinations," *J. Struct. Div. ASCE* 108:5 (1982) 978-997.

Ellingwood, B.R., and R.B. Corotis (1991), "Load Combinations for Buildings Exposed to Fires," *Engrg. J. AISC* 28:1 (1991) 37-44.

Ellingwood, B. (1998), "Reliability-based performance concept for building construction," in *Struct. Engrg. Worldwide*, Proc. Struct. Engrg. World Congress 1998, Elsevier, Paper T178-4 (CD-ROM).

Ellingwood, B. (2001), "Acceptable Risk Bases for the Design of Structures," *Progress in Struct. Engrg. and Materials* 3:2 (2001) 170-179.

*Ellingwood, B.R. (2005a), "Load Combination Requirements for Fire-Resistant Structural Design," *J.Fire Protection Engrg*. 15:1 (2005) 43-61.

*Ellingwood, B.R. (2005b), "Progressive Collapse-Resistant Design: Acceptable Risk Bases," in *Best Practices for Mitigating Risk of Progressive Collapse*, H.S. Lew, ed., Washington: National Institute of Building Sciences.

European Prestandard ENV 1991-2-7 (1998), "*Eurocode 1: Basis of design and actions on structures, Part 2-7: Accidental actions*," Brussels: Comite Europeen de Normalization 250.

Galambos, T.V., et al. (1982), "Probability-Based Load Criteria: Assessment of Current Design Practice," *J. Struct. Div. ASCE* 108:5 (1982) 959-977.

Gewain, R.G., and E.W.J. Troup (2001), "Restrained Fire Resistance Ratings in Structural Steel Buildings," *Engrg. Journal AISC* 38:2 (2001) 78-89.

Gustaferro, A.H., and L.D. Martin (1989), *PCI Design for Fire Resistance of Precast Prestressed Concrete*, 2nd ed., Chicago: Prestressed Concrete Institute.

Hamburger, R.O. (1996), "Implementing performance-based seismic design in structural engineering practice," *Proc. 11th World Conf. on Earthquake Engrg.*, Paper 2121, Elsevier Science Ltd.

Hurley, M. (2004), "Design Fire Scenarios," *Proc. NIST-SFPE Workshop for Development of a National R & D Roadmap for Fire Safety Design and Retrofit of Structures*, NISTIR 7133:107-117.

ICC (2006), *International Building Code, 2006 Edition,* Country Club Hills, Ill.: International Code Council.

ICC (2006), *ICC Performance Code for Buildings and Facilities*, Country Club Hills, Ill.: International Code Council.

ISO (1994), *ISO Standard 834: Fire resistance tests – Elements of building construction*, Geneva: International Organization for Standardization.

Jeanes, D.C. (1985), "Application of the Computer in Modeling Fire Endurance of Structural Steel Floor Systems," *Fire Safety Journal* 9 (1985) 119-135.

Kruppa, J. (2000), "Recent Developments in ire Design." *Progress in Struct. Engrg. and Materials* 2:1 (2000) 6-15.

Kumar, S., and V. Rao (1997), "Fire Loads in Office Buildings," *J. Struct. Engrg. ASCE* 3 (1997) 365-368.

*Lamont, S., B. Lane, A. Usmani, and D. Drysdale (2003), "Assessment of the Fire Resistance Test with Respect to Beams in Real Structures," *Engrg. J. AISC* 40:2 (2003) 63-75.

Lane, B. (2000), "Performance-Based Design for Fire Resistance," *Modern Steel Construction, AISC* (Dec. 2000) 54-61.

*Lie, T.T., ed. (1992), *Structural Fire Protection*, Manual of Engineering Practice No. 78, Reston, Va.: American Society of Civil Engineers.

Lie, T.T., and Almand, K.H., "A Method to Predict the Fire Resistance of Steel Building Columns," *Engrg. Journal AISC* 27 (1990) 158-167.

Liew, J.Y.R., and H. Chen (2004), "Explosion and Fire Analysis of Steel Frames Using Fiber Element Models," *J. Struct. Engrg. ASCE* 130:7 (2004) 991-1000.

Lim, L., A. Buchanan, P. Moss, and J.-M. Franssen (2004), "Numerical Modeling of Two-Way Reinforced Concrete Slabs in Fire," *Engrg. Struct.* 26:8 2004) 1081-1091.

May, P.J. (2004), "Making Choices About Earthquake Performance," *National Hazards Review* 5:2 (2004) 64–70.

*Meacham, B.J. (1997), "An Introduction to Performance-Based Fire Safety Analysis and Design with Applications to Fire Safety," in *Building to Last, Proc. Struct. Congress XV*, New York: American Society of Civil Engineers, 529-533.

Mehaffey, C., and T.Z. Harmathy (1984), "Failure Probabilities of Constructions Designed for Fire Resistance," *Fire and Materials* 8:2 (1984) 96-104.

*Milke, J.A. (1985), "Overview of Existing Analytical Methods for the Determination of Fire Resistance," *Fire Technology* 21:1 (1985) 59-65.

Milke, J.A. (2002), "Analytical Methods for Determining Fire Resistance of Steel Members," *SFPE Handbook of Fire Protection Engineering*, 3rd ed., DiNenno, ed., Quincy, Mass.: National Fire Protection Association.

Mowrer, F.W. (2003), "Overview of Performance-Based Fire Protection Design," *Fire Protection Handbook*, 19th ed., Section 3, Chapter 14, Quincy, Mass.: National Fire Protection Association.

Mowrer, F.W. (2004), "Performance—The Sine Qua Non," in *Fire Protection of Structural Steel in High-Rise Buildings*, Civil Engineering Research Foundation, pp. 28-50.

NFPA (2006), *Building Construction and Safety Code, 2006 Edition*, NFPA 5000, Quincy, Mass.: National Fire Protection Association.

National Research Council of Canada (1996), *National Building Code of Canada, 1995*, Ottawa: National Research Council of Canada.

Newman, G.M., J.T. Robinson, and C.G. Bailey (2000), *Fire safe design: a new approach for multi-storey steel-framed buildings*, Ascot, U.K.: Steel Construction Institute.

Nuclear Regulatory Commission, (NUREG-1824, Vols. 1-7),Verification and Validation of Selected Fire Models for Nuclear Power Plant Applications, U.S. Office of Nuclear Regulatory Research, Rockville, MD., May 2007

Pate-Cornell, E. (1994), "Quantitative Safety Goals for Risk Management of Industrial Facilities," *Struct. Safety* 13:3 (1994) 145-157.

Ruddy, J.L., J.P. Mario, S.A. Ioannides, and F. Alfawakhiri (2003), *Fire Resistance of Structural Steel Framing*, Steel Design Guide 19, Chicago: American Institute of Steel Construction.

*SFPE (2000), *Guide to Performance-Based Fire Protection Analysis and Design of Buildings*, Bethesda, Md.: Society of Fire Protection Engineers.

*SFPE/SEI (2003), *Designing Structures for Fire*, published for the Society of Fire Protection Engineers by DesTech Publications, Inc.

*SFPE (2004), *Engineering Guide: Fire Exposures to Structural Elements*, Bethesda, Md.: Society of Fire Protection Engineers.

SFPE (2008), *SFPE Handbook of Fire Protection Engineering*, Bethesda, Md: Society of Fire Protection Engineers.

Siu, J. (2004), "Challenges Facing Engineered Structural Fire Safety—A Code Official's Perspective," *Proc. NIST-SFPE Workshop for Development of a National R & D Roadmap for Fire Safety Design and Retrofit of Structures*, NISTIR 7133:61-69.

SSRC Guide to Stability Design Criteria for Metal Structures (1998), 5th ed., T.V. Galambos, ed., New York: John Wiley.

Stewart, M.G., and R.E. Melchers (1997), *Probabilistic risk assessment of engineering systems*, London: Chapman & Hall.

Chapter 4

Design of Concrete Structures

James Milke, Ph.D., P.E., University of Maryland
Stephen Pessiki, Ph.D., Lehigh University
Long Phan, Ph.D., P.E., National Institute of Standard and Technology

4.1 INTRODUCTION

Concrete construction may serve as a fire-resistant barrier and/or as a load-bearing component of a structure. Thus, an analysis of the fire resistance of concrete construction may need to assess the characteristics of the construction relative to one or both of these functions. In particular:

- An assembly serving as a barrier to prevent the spread of fire, such as a floor-ceiling assembly or a wall, contributes to the fire protection strategy of compartmentation. As such, the barrier needs to be a sufficient insulator, limiting heat transmission to the unexposed side of the barrier. This is assessed by adopting limits for temperature rise on the surface of the unexposed side of the barrier. ASTM E119 (ASTM 2007) limits the average temperature rise above ambient on the unexposed side to 139 °C.

   The fire resistance of an assembly serving as a barrier is assessed via a heat transfer analysis. The purpose of the heat transfer analysis is to confirm that the temperature rise on the unexposed side of the assembly resulting from the fire exposure meets established limits. The heat transfer analysis requires the dimensions and geometry of the assembly and the thermal material properties (see Section 4.2). In addition, the fire exposure needs to be characterized (see Chapters 2 and 3).

- A load-bearing assembly needs to maintain its structural integrity, supporting the weight of the structure and its contents, despite the effects of the fire exposure. To achieve this aspect of performance, the load-carrying ability of the structural element needs to be preserved, in spite of changes in mechanical properties associated with an increase in temperature of the structural element and the imposition of thermal strain.

   A fire resistance assessment of load-bearing assemblies requires both a heat transfer and structural analysis. The heat transfer analysis requires information similar to that for a barrier, but now is conducted to assess temperatures of reinforcing elements or a temperature distribution within the assembly. Based on this information, a structural analysis is conducted that assesses load-carrying ability or structural stability, accounting for changes in material properties, spalling, and induced thermal strains.

Concrete columns are an example of an assembly that needs to satisfy only the structural integrity criterion. In contrast, walls and floor–ceiling assemblies are examples of barriers that are also load bearing and thus would need to satisfy both considerations in order to be considered fire resistant.

The design of concrete construction for fire resistance can be based on engineering principles incorporated into calculation procedures described in this chapter (Milke 1999). A variety of procedures are available to estimate the fire resistance of concrete flexural members and concrete columns. These procedures range from the application of algebraic equations to finite element computer models.

The calculation methods included in this chapter are based primarily on research conducted by the Portland Cement Association (PCA). Much of the research was sponsored by the Precast/ Prestressed Concrete Institute (PCI) and the PCA (specific references from this research are cited throughout this chapter). This chapter summarizes the technical basis for the calculation methods, presents the methods, and provides examples to illustrate their application. The information presented in this chapter is drawn from ACI 216.1-07, the 6th edition of the *PCI Design Handbook* (PCI 2004); PCI manual, MNL-124-89, *Design for Fire Resistance of Precast Prestressed Concrete* (PCI 1989); the Concrete Reinforcing Steel Institute manual, *Reinforced Concrete Fire Resistance* (CRSI 1980), as well as results of recent studies. These references are recognized as the principal references in the field and for years were recognized by the model codes as acceptable resource documents for determining fire resistance ratings of concrete construction by other than prescriptive means.

4.1.1 Scope

Concrete buildings exist in many forms. For example, concrete buildings may be cast-in-place, precast on site (tilt-up construction), or precast in a manufacturing facility. Concrete buildings may be made with mild steel reinforcement or prestressing steel reinforcement. For prestressed concrete, the construction may be pretensioned or post-tensioned, and bonded or unbonded. The methods described in this chapter apply to this wide range of types of concrete structural members and assemblies. For global structural analysis of entire concrete building in fire, advanced calculation models, developed based on acknowledged engineering principles and assumptions of the theory of structural mechanics, as generally described in the Eurocode as well as in section 3.4.3 of this report, may be used with the effects of thermal expansions and large deformations, degradation of concrete material properties and potential concrete spalling as prescribed in section 4.2, appropriately taken into account.

4.1.2 Technical Basis for Calculating Fire Resistance of Concrete Construction

Current practice in the design of concrete structures for fire is based on hundreds of fire tests of cast-in-place and precast concrete assemblies. These fire tests have been performed on a variety of structural assemblies including beams, slabs, columns, and walls.

Reports of a number of tests of flexural elements sponsored by PCI have been issued by Underwriters Laboratories (UL). Most of the reports have been reprinted by PCI, and the results of the tests are the basis for UL's listings and specifications for non-proprietary products such as double tee beams and hollow-core slabs. PCA has also conducted many fire tests of precast flexural elements, including simply supported and continuous slabs and beams. Test results published as Research and Development Bulletins available from PCA also have been the basis for papers published elsewhere.

Fire tests of reinforced concrete columns have been conducted by PCA and the National Research Council of Canada. While no tests are known to have been conducted for prestressed concrete columns, results from tests of reinforced columns are considered to be equally applicable to prestressed concrete columns with adjustment made for the difference in thermal properties between mild reinforcing steel and prestressing strand as appropriate. Concrete cover requirements are less for precast concrete assemblies manufactured under controlled conditions in a precast plant. This may also influence the relative behavior of cast-in-place columns as compared with precast concrete columns, as discussed more fully later in the chapter.

The fire resistance of concrete walls is normally governed by the ASTM E119 criteria for temperature rise of the unexposed surface rather than by structural behavior during fire tests. This is due to the low level of stresses under service load, even in concrete bearing walls. In most cases, the concrete cover for steel reinforcement required by code exceeds that required for fire protection so there is, in effect, reserve structural fire resistance within the concrete wall.

Although many of the tests performed over the years were conducted to determine fire ratings for specific structural elements, many of the tests were performed in conjunction with broad research studies whose objectives have been to understand the behavior of concrete elements subjected to fire. The knowledge gained from these tests has resulted in the development of listings of fire-resistive concrete building assemblies and procedures for determining the fire resistance of precast concrete members by calculation.

The remainder of this chapter reviews material properties, including their dependence on temperature, and computational methods for heat transfer and structural analyses of reinforced and prestressed concrete assemblies exposed to fire.

4.2 MATERIAL PROPERTIES

This section on material properties of concrete at elevated temperatures is divided into two major sections. The first section presents information on material properties for normal strength concrete (NSC). The second section presents information on material properties for high strength concrete (HSC). In these guidelines, NSC is defined as concrete with a compressive strength less than 83 MPa (12,000 psi), and HSC is defined as concrete with a compressive strength greater than 83 MPa.

The concrete material properties required for thermal and structural calculations are thermal and mechanical properties. The thermal properties described include thermal conductivity, volumetric specific heat, and density. The mechanical properties include concrete compressive strength, modulus of elasticity, shear modulus, tensile strength, and coefficient of thermal expansion. These properties are required for engineering analyses in typical design applications.

Additional properties may also be required in more advanced analyses. These include:

- Porosity
- Moisture content
- Poisson's ratio
- Creep parameters

In some cases, multiple sources of data for material properties are provided. These are included to demonstrate the range in property values that have been determined, where the variations are attributable primarily to variations in the experimental procedure used to determine the properties. In general, standard experimental procedures are not available to assess the noted material properties over the range of temperatures of interest for concrete.

4.2.1 Normal Strength Concrete

4.2.1.1 Thermal Properties

At a minimum, an unsteady conduction heat transfer analysis requires the following material properties to be known:

- Thermal conductivity, λ (W/m·K)
- Specific heat, C_p (J/kg·K)
- Density, ρ (kg/m^3)

The reported thermal properties of concrete are obtained from small-scale tests. Full-scale members are likely to perform differently, especially with regard to moisture migration. Therefore, these properties should be considered to be useful only for developing estimates of the thermal response of a concrete structural member.

4.2.1.1.1 Thermal Conductivity

Thermal conductivity, λ, of concrete is primarily a function of concrete densities and aggregate type. Values of λ for concrete with density ranging from 800 kg/m^3 to 2,400 kg/m^3 are listed in ACI 216.1 (ACI 2007). For simple analyses, the following averaged values may be used for the thermal conductivity of NSC with normal weight and light weight aggregate (EC2 2002):

Normal weight aggregate NSC: $\lambda = 1.3$ W/m·K
Lightweight aggregate NSC: $\lambda = 0.5$ W/m·K

For more comprehensive analysis where temperature-dependent thermal conductivity is needed, correlations are provided in Appendix A of this chapter (Section 4.6) and are also graphically depicted in Figure 4.1 (Harmathy 1970a; Harmathy and Allen, 1973). The thermal conductivity of normal weight aggregate concrete is greater than that of lightweight aggregate concrete, which makes lightweight concretes better insulators than normal weight concretes. This affects the heat transmission aspect of fire resistance for normal weight and lightweight aggregate concrete slabs and walls as well as the temperature rise of any reinforcement within the concrete (for concrete walls, slabs, and columns).

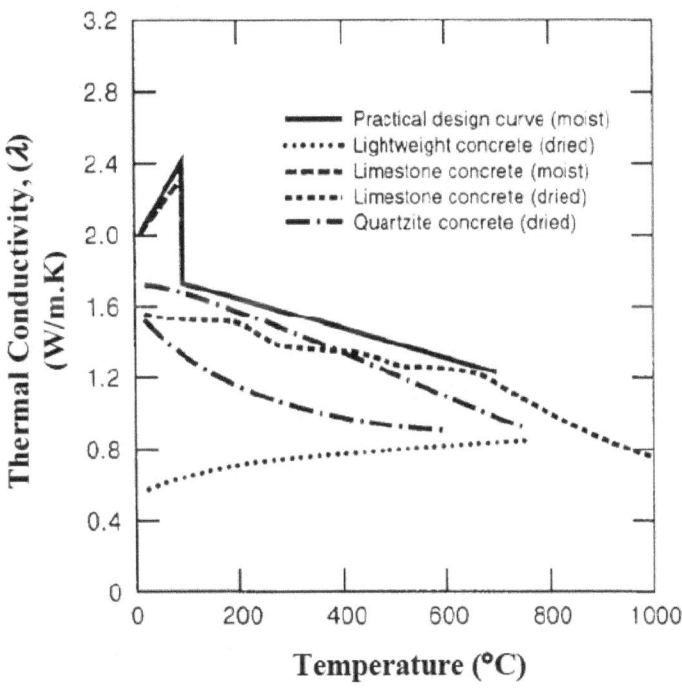

FIGURE 4.1. Thermal Conductivity of Different Structural Concretes (Schneider, 1988)

4.2.1.1.2 Volumetric Specific Heat

Volumetric specific heat, $\rho.C_p$,   For simple analyses, the following values may be used for the volumetric specific heat of concrete (EC2 2002):

Normal weight concrete: $\rho.C_p = 2.6 \, \text{MJ/m}^3 \cdot \text{K}$
Lightweight concrete: $\rho.C_p = 1.5 \, \text{MJ/m}^3 \cdot \text{K}$

The volumetric specific heat of concrete at elevated temperatures is illustrated in Figure 4.2 (Harmathy 1970b). Again, the volumetric specific heat differs for the two aggregates, with the normal weight aggregate concretes having a greater volumetric specific heat than the lightweight aggregate concretes. As with thermal conductivity, this will affect the temperature rise within and on the unexposed surface of the concrete member. Correlations to determine the volumetric specific heat at selected temperatures are included in Appendix A (Section 4.6 below).

4.2.1.1.3 Density

Concrete density, ρ, does not vary significantly over the temperature range typically associated with common fires. In general, concrete density decreases with increasing temperature. This is due mostly to mass loss incurred in the heat-induced release and evaporation of free and chemically-bound waters in unsealed concrete. For calculation purpose, the densities of structural concretes commonly used in concrete building construction can be taken as 1760 kg/m³ (110 lb/ft³) for light weight aggregate concrete, and 2330 kg/m³ (145 pcf) for normal weight aggregate concrete. Lightweight aggregate concrete is more commonly used in precast concrete construction for elements such as double tee roof framing members. Normal weight aggregate concrete is most commonly used in cast-in-place construction.

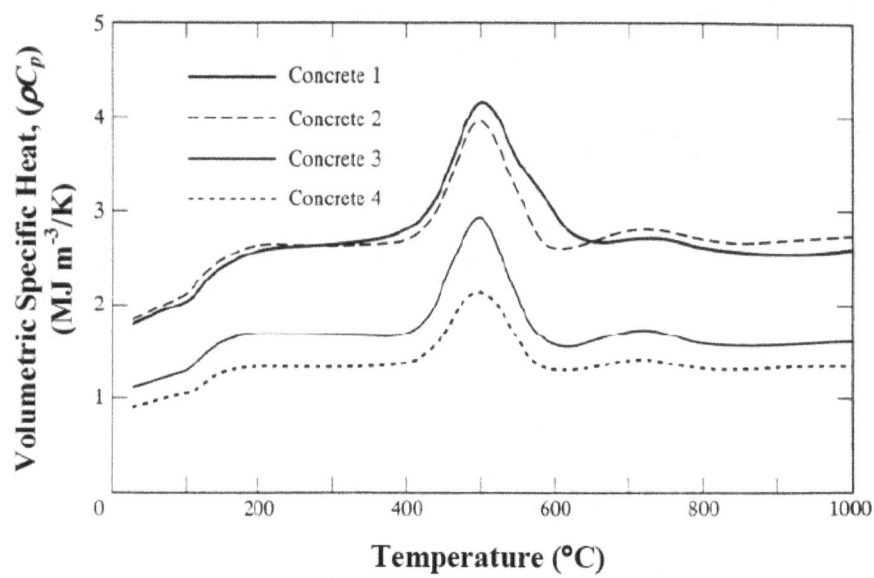

FIGURE 4.2. Volumetric Specific Heat of Normal Weight (Concretes 1 and 2) and Lightweight Aggregate (Concretes 3 and 4) Concretes (Harmathy 1970b)

4.2.1.2 Mechanical Properties

Concrete temperature-dependent mechanical properties, including compressive and tensile strength, modulus of elasticity, shear modulus, and coefficient of thermal expansion are discussed in this section.

4.2.1.2.1 Compressive Strength

The typical compressive strength value, f'_c, specified for design is the 28-day compressive strength. Prestressed construction also includes a requirement for the compressive strength of the concrete at the time of load transfer due to prestressing. The requirement for high early strength to allow the load transfer to occur soon after concrete placement (8 to 12 hours) often leads to actual 28-day compressive strengths that significantly exceed the specified design value.

Stress–strain curves for normal and lightweight aggregate concretes at elevated temperatures are provided in Figure 4.3 (Harmathy 1993). The slope of the curve decreases with increasing temperature, indicating a decline in the modulus of elasticity with temperature, as is confirmed in subsequent figures in this section.

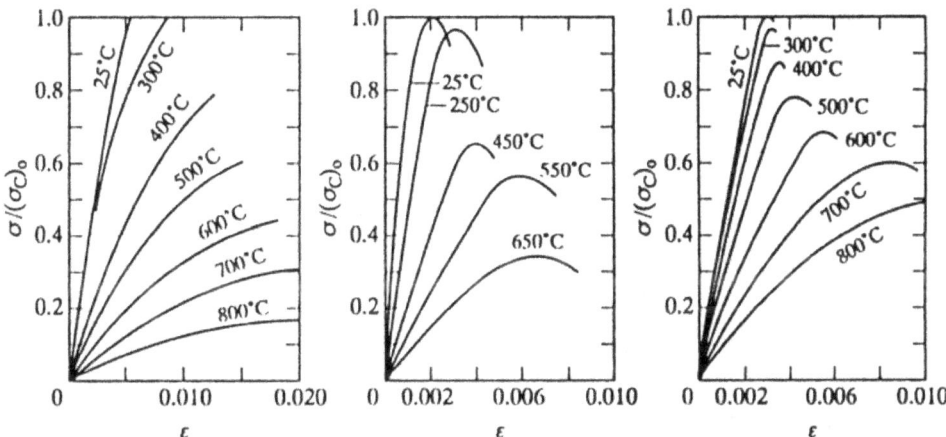

FIGURE 4.3. Stress–Strain Curves at Elevated Temperature for Normal and Lightweight Aggregate Concretes (Harmathy 1993)

Concrete compressive strength varies differently at elevated temperatures depending on the stress and heating conditions. In laboratory tests, concrete compressive strength at elevated temperature are typically evaluated using three different test methods, called *stressed*, *unstressed*, and *unstressed residual property* test method. Each test method yields results with different application.

1. *Stressed* test method: Apply a constant load to the test sample (ranging from 25% to 55 % of ultimate room-temperature compressive strength), heat the sample to its target temperature, allow time for steady state temperature condition to be achieved within the test sample (near uniform temperature), and increase the load to failure. Mechanical properties determined using this method are applicable for use in calculating structural response of concrete members whose main function is to carry a large portion of compressive load, such as columns. Figure 4.4 shows the variation with respect to temperature of compressive strength obtained under this test method for normal weight and lightweight aggregate NSC (Phan, 2002).

2. *Unstressed* test method: Heat the unloaded sample to its target temperature, allow time for near uniform temperature to be achieved within the test sample, and increase the load to failure. Mechanical properties determined using this method are applicable for use in calculating structural response of concrete members whose main function is to serve as barrier and carry only minor compressive load, such as partition walls or beams. Figure 4.5 shows variation of compressive strength of NSC obtained by various studies under the unstressed test method (Phan, 2002)

3. *Unstressed residual property* test method: Heat the unloaded sample to its target temperature, allow the sample to cool to room temperature, and increase the load to failure. This information is particularly relevant to post-fire integrity assessments of concrete construction. Figure 4.6 shows variation of compressive strength of NSC obtained by various studies under the unstressed residual properties test method (Phan, 2002)

Designers should refer to properties determined from the test method that most closely relates to the application being considered, e.g., in a design application, properties of loaded samples (stressed test method) should be considered. In contrast, a post-fire integrity analysis should consider the data from tests where the sample was heated and then allowed to cool prior to the application of a load (unstressed residual properties test method).

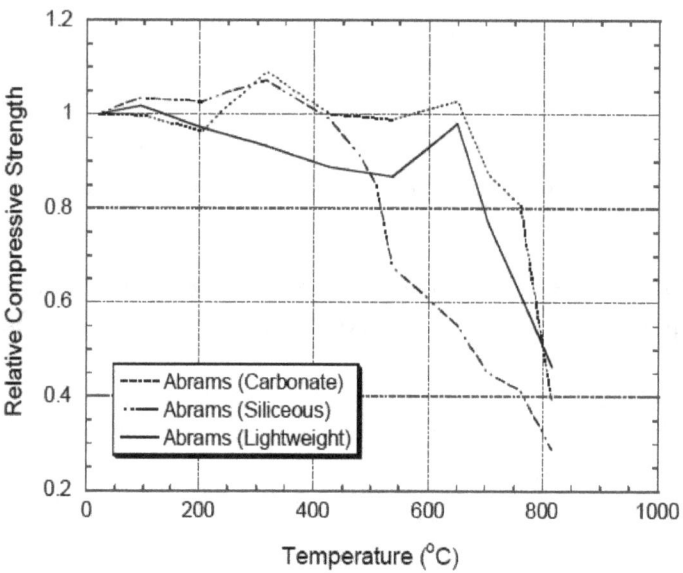

FIGURE 4.4. Relative Compressive Strength of NSC (normal and lightweight aggregate) under Stressed Test Condition

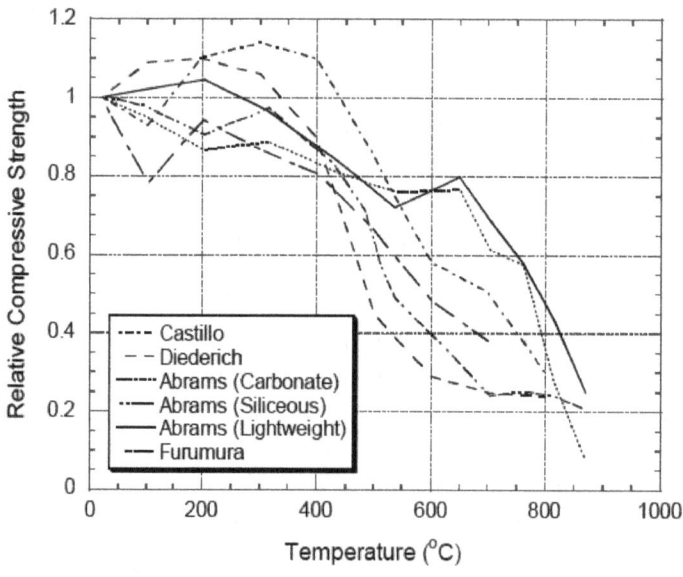

FIGURE 4.5. Relative Compressive Strength of NSC (normal and lightweight aggregate) under Unstressed Test Condition

68

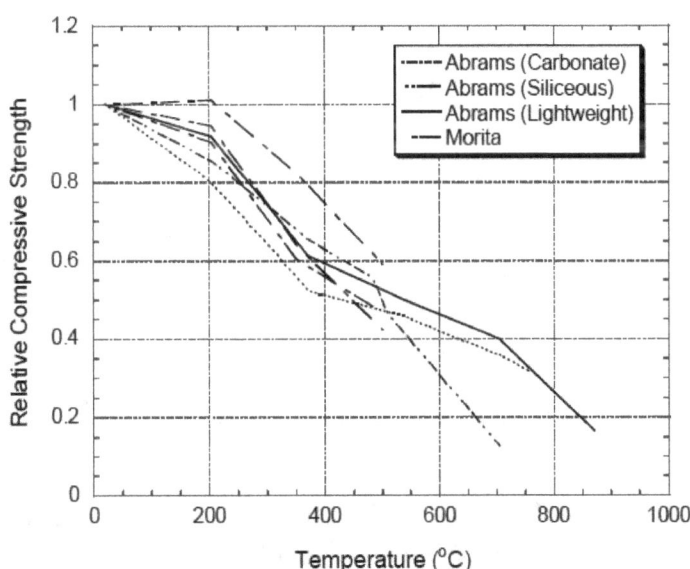

FIGURE 4.6 Relative Compressive Strength of NSC (normal and lightweight aggregate) under Unstressed Residual Strength Test Condition

The following correlations, proposed by Lie (1992), provide estimates of the compressive strength at elevated temperature for NSC made with siliceous, carbonate, or expanded shale aggregate. These estimates are conservative and can be used to determine NSC compressive strength in calculation methods where compressive strength at elevated temperature is required.

$$f_{c,T} = f_{c,0} \qquad\qquad \text{for T} \le 450\ {}^\circ\text{C} \qquad\qquad (4.1)$$

$$f_{c,T} = f_{c,0}\left[2.011 - 2.353\frac{T-20}{1000}\right] \qquad \text{for T} > 450{}^\circ\text{C} \qquad (4.2)$$

4.2.1.2.2 *Modulus of Elasticity*

The variation in the modulus of elasticity, E_c of NSC over a range of temperatures is provided in Figure 4.7 (Cruz 1966). Similar reductions in the elastic modulus as a function of temperature are evident for both the lightweight aggregate concrete and the normal weight concrete.

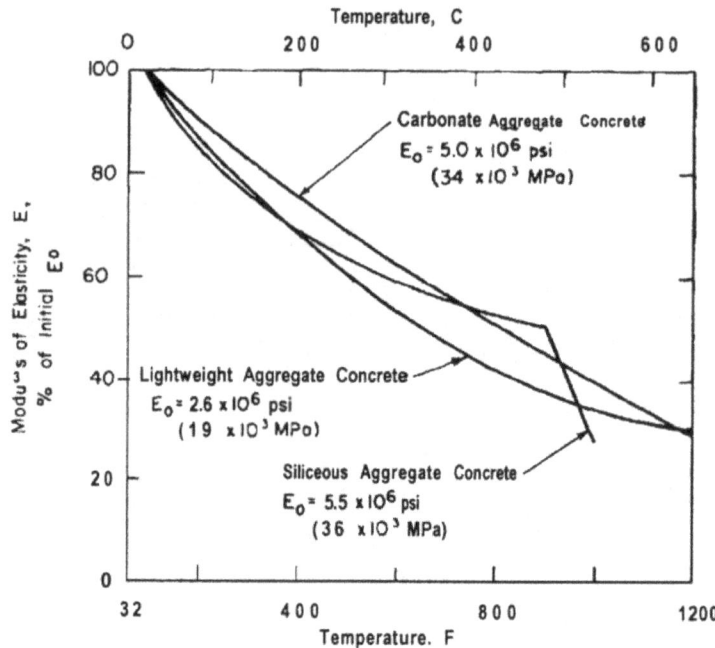

FIGURE 4.7. Modulus of Elasticity of NSC at Elevated Temperatures (Cruz 1966)

4.2.1.2.3 Shear Modulus

The shear modulus at elevated temperatures is presented in Figure 4.8.

4.2.1.2.4 Tensile Strength

The tensile strength of concrete is often empirically expressed as a product of a multiplier, γ, and the square root of the compressive strength, where γ depends on the concrete density and the type of test. The American Concrete Institute (ACI 318 1989) suggests the following relationship for the two strengths at room temperature:

$$f_t = 620\sqrt{f'_c} \qquad\qquad (4.3)$$

Where:
f'_c = Compressive ultimate strength (Pa)
f_t = Tensile ultimate strength (Pa)

70

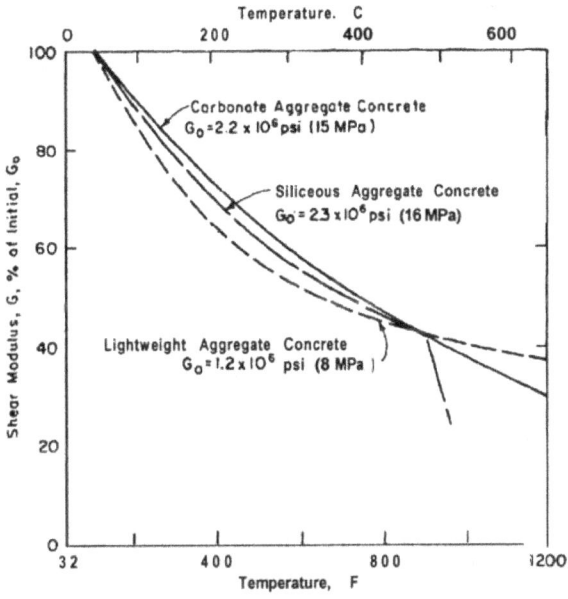

FIGURE 4.8. Shear Modulus of Concrete at Elevated Temperatures (ACI 216 1989)

The tensile strength of concrete is relatively small, typically about 10% of the compressive strength, and is usually neglected in practical design applications.

4.2.1.2.5 *Coefficient of Thermal Expansion*

The coefficient of thermal expansion is used to predict thermally induced loads and curvatures in a structure. The coefficient of thermal expansion of concrete was measured by Cruz at elevated temperatures (see Figure 4.9). Correlations to fit the data by Cruz were provided by Lie (1992) and are provided in Section 4.6 (Appendix A).

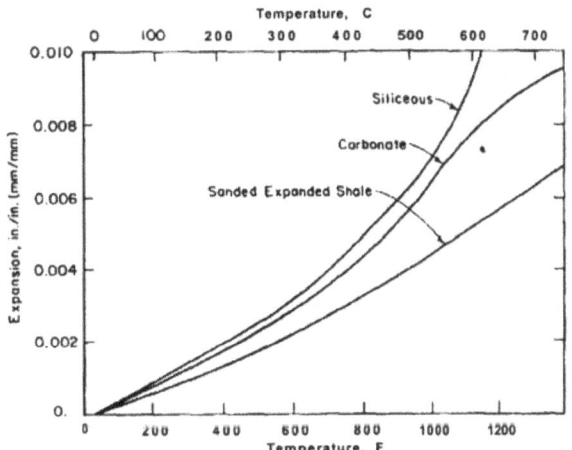

FIGURE 4.9. Coefficient of Thermal Expansion of Concrete (ACI Committee 216 2007)

71

4.2.2 High Strength Concrete

HSC is more widely used in columns of cast-in-place construc-tion or in heavily loaded flexural elements such as precast prestressed inverted tee girders. In pretensioned construction, the need for a sufficiently large compressive strength at early age to allow the transfer of prestress often leads to an actual 28-day compressive strength that exceeds the specified 28-day compressive strength.

4.2.2.1 Thermal Properties

Studies (Van Geem et al., 1996; Burg and Ost, 1994) have shown high strength concrete and normal strength concrete has very similar thermal properties. Thus thermal properties provided in section 4.2.1.1 can also be used for analysis of high strength concrete.

4.2.2.2 Mechanical Properties

4.2.2.2.1 Compressive Strength

As with the data for the compressive strength of NSC, the compressive strength for HSC also varies differently at elevated temperature depending on the stress and temperature levels. The data for compressive strength is reflected in Figures 4.10 to 4.12 (Phan, 2003, 2007). Phan proposes that, for design purposes, the compressive strength of HSC be selected from Figure 4.13 (Phan, 2003, 2007).

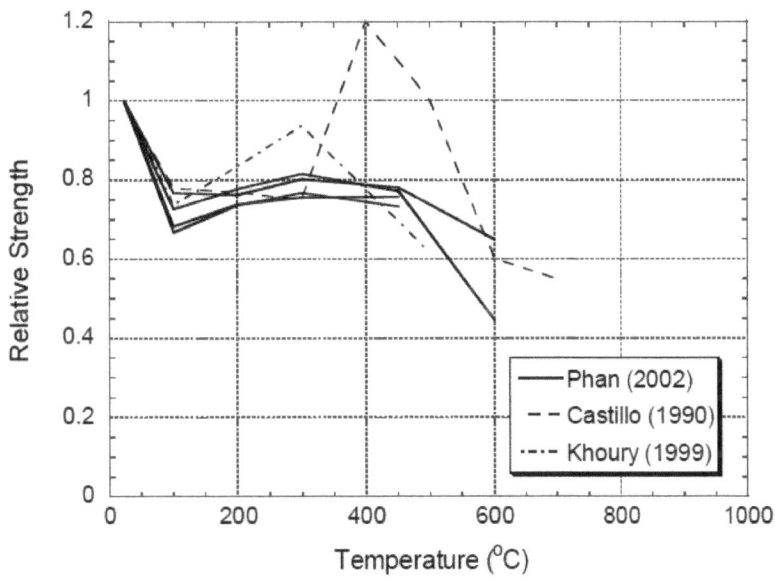

**FIGURE 4.10. Relative Compressive Strength of HSC
under the Stressed Test Method (Phan, 2003, 2007)**

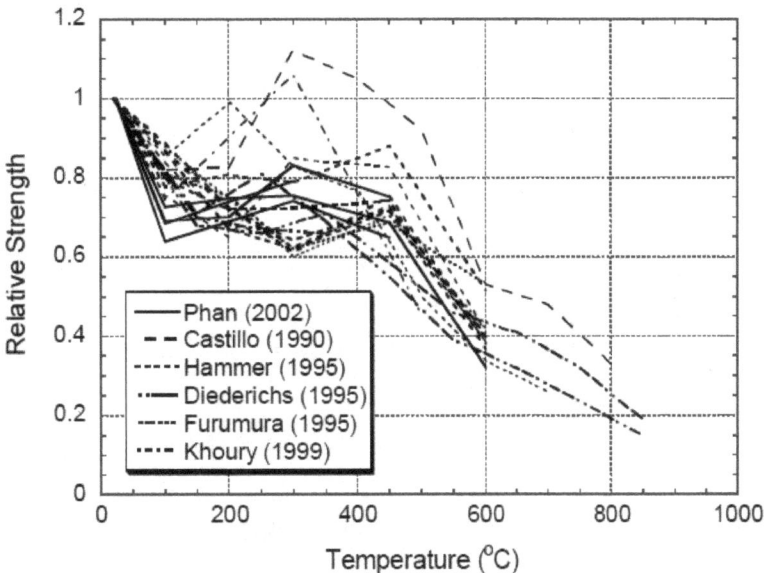

**FIGURE 4.11. Relative Compressive Strength of HSC
under the Unstressed Test Method (Phan, 2003, 2007)**

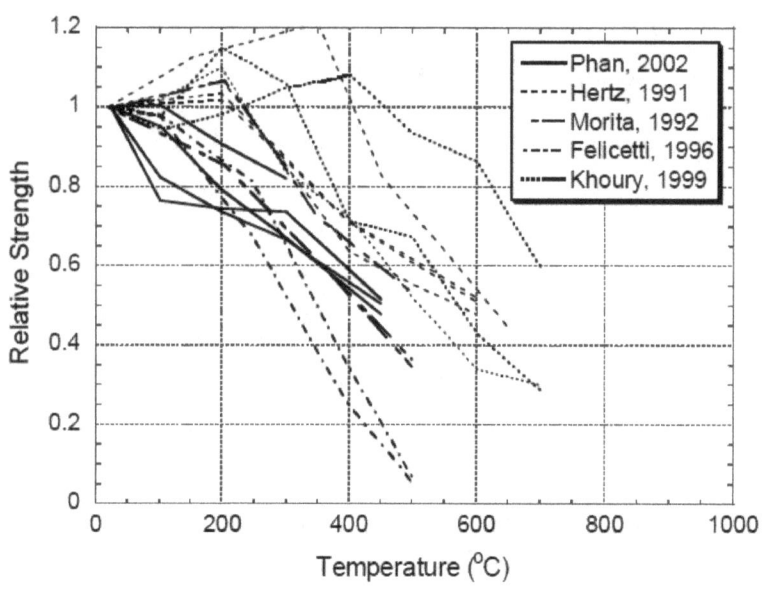

FIGURE 4.12. Relative Compressive Strength of HSC under Unstressed Residual Property Test Method (Phan, 2003, 2007)

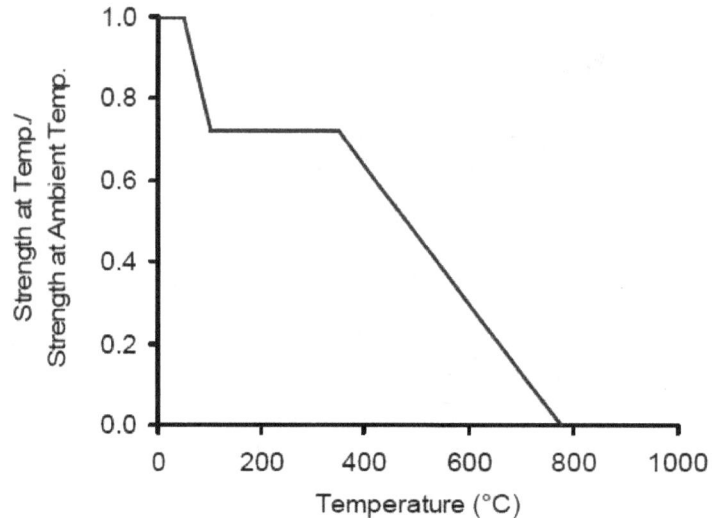

FIGURE 4.13. Relative Compressive Strength of HSC for Design Applications (Phan, 2003)

Correlations to describe the line segments in Figure 4.13 are:

$$f_c = f_{c,20} \qquad\qquad\qquad\qquad\qquad T \leq 50\ °C$$
$$f_c = f_{c,20}\left[1.28 - .0056T\right] \qquad\quad 50\ °C < T \leq 100\ °C$$
$$f_c = 0.72 f_{c,20} \qquad\qquad\qquad\quad 100\ °C < T \leq 350\ °C$$
$$f_c = f_{c,20}\left[1.31 - .00168T\right] \qquad 350\ °C < T \leq 778\ °C$$
$$f_c = 0 \qquad\qquad\qquad\qquad\qquad\quad T > 778\ °C$$

4.2.2.2.2 Modulus of Elasticity

The variation in the modulus of elasticity, E_c, of HSC made of limestone coarse aggregate over a range of temperature is provided in Figure 4.14 (Phan, 2001). The data were obtained under three different test methods (see Section 4.2.1.2.1) and show that the variation of HSC's modulus of elasticity is not significantly influenced by the stress the stress level during heating. In addition, the variation of HSC's modulus of elasticity at elevated temperature in general follows the trend observed for NSC's modulus of elasticity (see Figure 4.7).

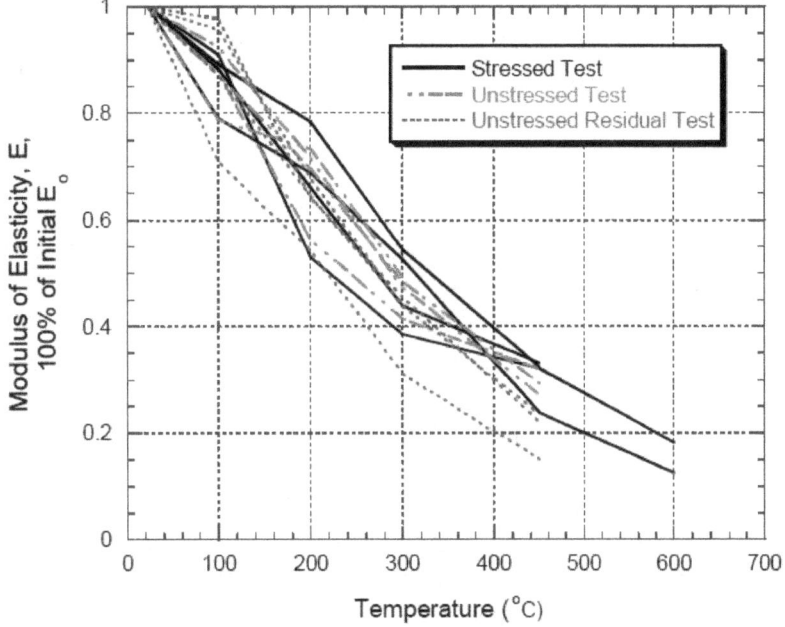

FIGURE 4.14. Modulus of Elasticity of HSC at Elevated Temperature (Phan, 2001)

75

4.2.3 Steel Reinforcement

Both cast-in-place and precast concrete building structures are reinforced in a variety of ways. Types of reinforcement include mild steel reinforcement bars, wire mesh, high-strength prestressing wire, and strand and high-strength bars. The discussion in this section applies to all types of steel reinforcement commonly used in cast-in-place and precast concrete building structures. For structural steel, discussion is provided in Chapter 5.

Heat transfer through steel reinforcement is usually neglected given the relatively small amount of steel mass in a concrete assembly. As such, thermal properties of steel reinforcement are usually unnecessary in design applications. However, there are cases where it may be necessary to include the presence of the steel reinforcement in the thermal analysis. One example may be a heavily reinforced element of relatively small overall dimensions, such as a concrete corbel. A second example may be a detail that includes embedded steel hardware such as plates or angles welded to the main reinforcement in a member. The steel hardware may be exposed and thus provide a direct path to conduct heat to the reinforcement. In this situation, the temperature of the reinforcement may not be the same as the surrounding concrete, or the steel may more readily conduct heat into the concrete, thereby necessitating that the presence of the steel be considered in the analysis.

4.2.3.1 Thermal Properties

The thermal properties of steel reinforcement may be considered to be the same as for structural steel (see Chapter 5).

4.2.3.2 Mechanical Properties

Variations in mechanical properties of steel reinforcement, including yield strength f_y and modulus of elasticity E_y at elevated temperatures are shown in Figures 4.15 and 4.16.

4.3 THERMAL ANALYSIS

There are two purposes for conducting thermal (heat transfer) analyses of fire-exposed concrete assemblies:

1. Assess heat transmission to the unexposed side, i.e., compare the temperature rise on the surface of the unexposed side to performance criteria.
2. Provide input to structural analysis, i.e., determine the temperature of the steel reinforcements or temperature distribution through the assembly.

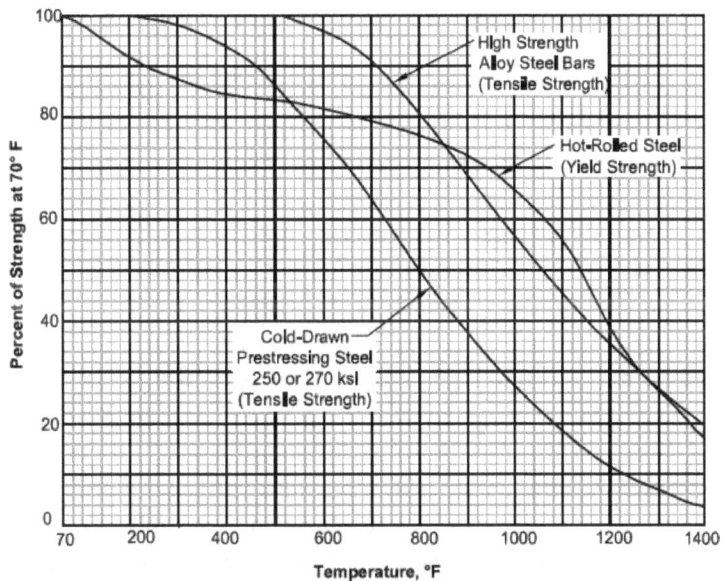

FIGURE 4.15. Strength–Temperature Relationships for Various Steel Reinforcements (ACI 216.1-07)

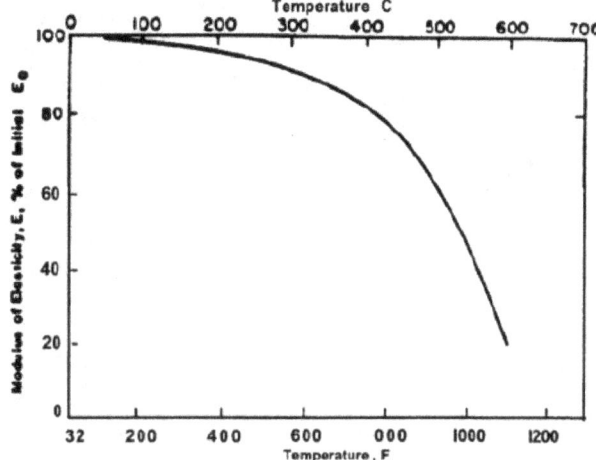

FIGURE 4.16. Modulus of Elasticity–Temperature Relationships for Steel (Weigler and Fischer 1964)

4.3.1 ASTM E119 Standard Fire Exposure

Heat transfer through concrete members exposed to the conditions associated with the ASTM E119 standard fire exposure has been studied by numerous individuals. These studies have resulted in:

- Tables to identify the minimum thickness of concrete required to limit the temperature rise on the surface of the unexposed side to acceptable limits (according to ASTM E119). These tables form the basis of the equivalent thickness method in ASCE/SFPE 29.

- Graphical methods to assess temperature rise at specific locations of interest within the concrete, i.e., at the location of reinforcement elements.
- Computer algorithms to assess temperature rise in reinforced concrete columns.

Numerous graphical solutions of the temperature distribution within fire-exposed structural members have been presented in the literature. However, most of the graphical solutions are limited to cases involving the standard ASTM E119 exposure.

Graphs of the type presented in Figures 4.17 and 4.18 provide the temperature distribution within concrete slabs based on data from tests with the ASTM E119 standard fire exposure (Abrams and Gustaferro 1968). Graphs are provided for three aggregates: siliceous, carbonate and sand-lightweight. These graphs are applicable for determining the temperature at a particular depth into a concrete slab, often at the location of reinforcing or pre-stressing steel within the slab. Additional graphs are available to determine the temperature rise on the unexposed surface of concrete slabs, beams, and columns exposed to the ASTM E119 test as a function of thickness and aggregate (ACI 216.1-07.) According to the review by Hosser et al. (1994), the graphical thermal analysis procedure included in the PCI guide (1989) provides good agreement with actual test data.

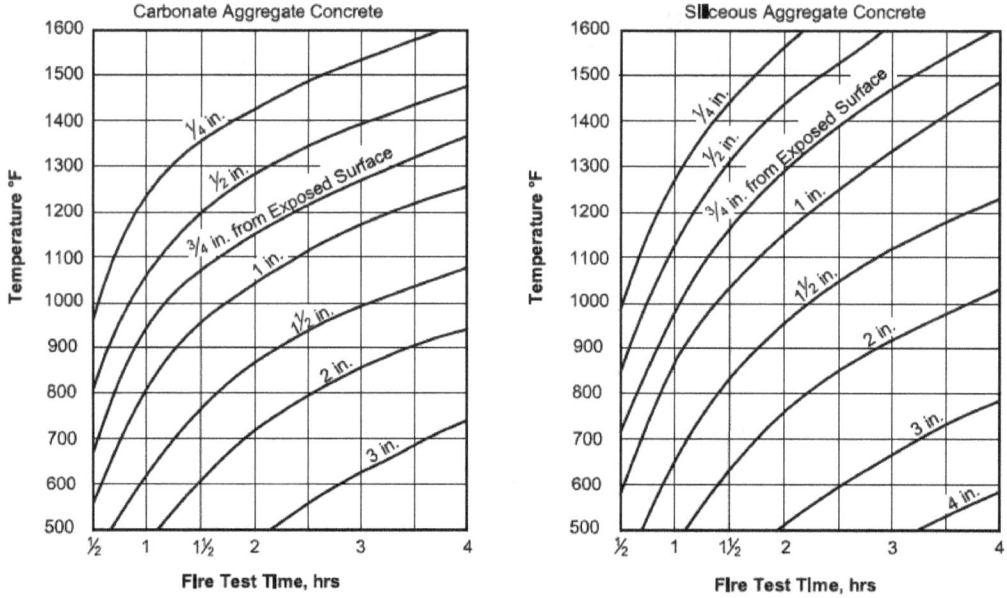

FIGURE 4.17. Temperatures Within Normal Weight Concrete Slabs or Panels During ASTM E119 Exposure (Abrams and Gustaferro 1968)

78

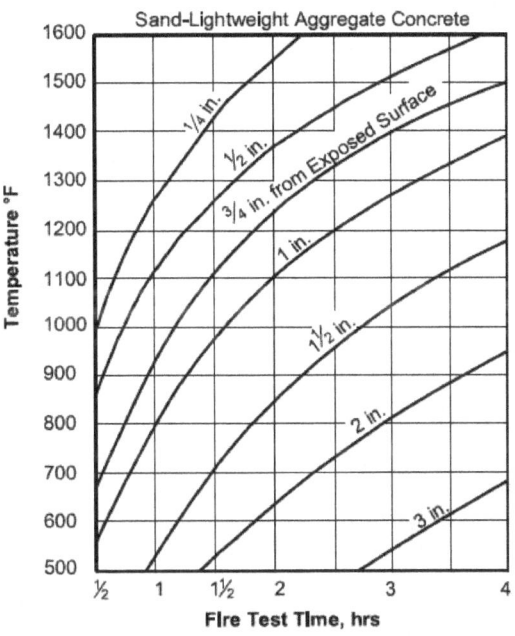

FIGURE 4.18. Temperatures Within Sand–Lightweight Concrete Slabs or Panels During ASTM E119 Exposure (Abrams and Gustaferro 1968)

Similarly, graphical solutions are presented for stemmed units exposed to the standard ASTM E119 fire exposure. Examples are presented in Figures 4.19 to 4.22.

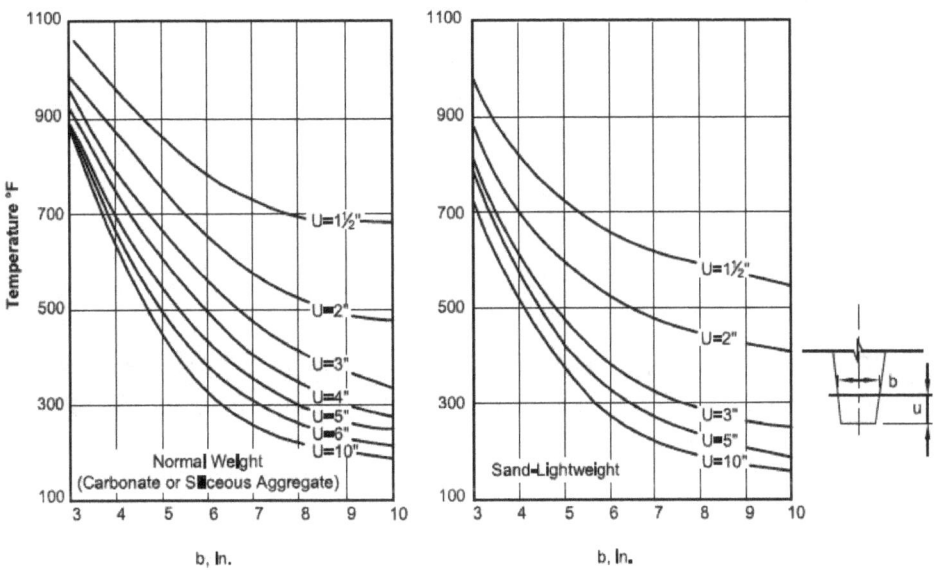

FIGURE 4.19. Temperatures on Vertical Centerline of Stemmed Units at 1 Hour of ASTM E119 Exposure (Ehm and van Postel 1967)

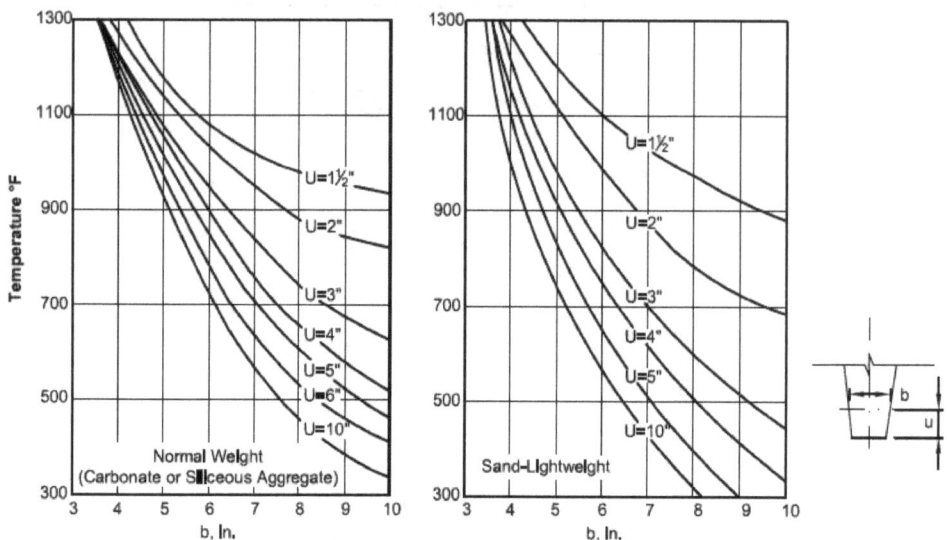

FIGURE 4.20. Temperatures on Vertical Centerline of Stemmed Units at 2 Hours of ASTM E119 Exposure (Ehm and van Postel 1967)

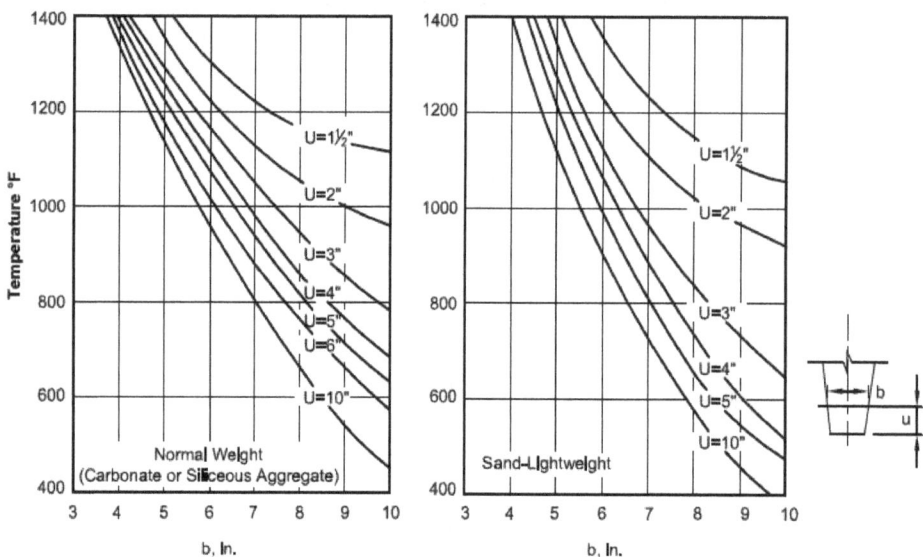

FIGURE 4.21. Temperatures on Vertical Centerline of Stemmed Units at 3 Hours of ASTM E119 Exposure (Ehm and van Postel 1967)

80

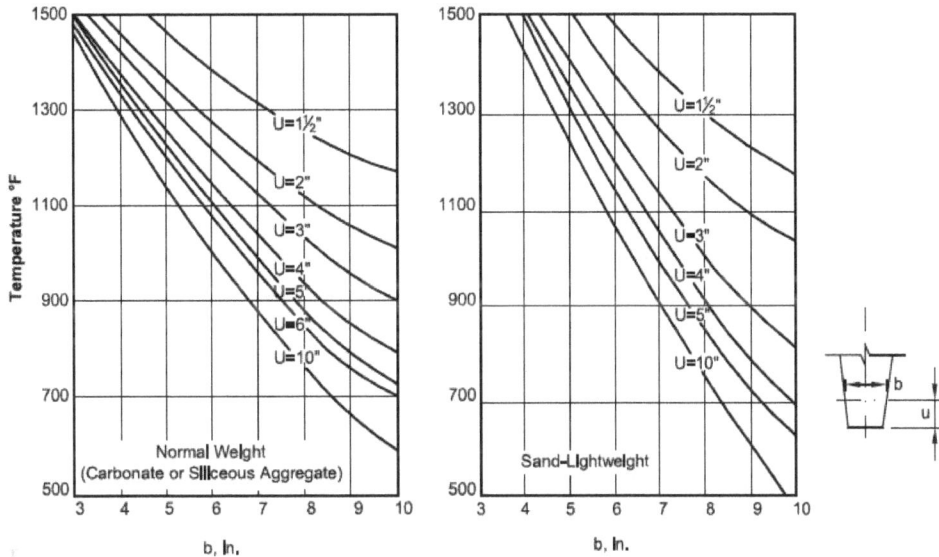

FIGURE 4.22. Temperatures on Vertical Centerline of Stemmed Units at 4 Hours of ASTM E119 Exposure (Ehm and van Postel 1967)

Lie (1972) provided a series of graphs for one-dimensional analyses of the temperature distribution in walls or slabs exposed on one or two sides to the ASTM E119 exposure as illustrated in Figure 4.23. The graphs can also be used for two-dimensional assemblies such as columns or beams by applying the principle of superposition. Because the material properties are assumed to be temperature-dependent, average properties need to be identified.

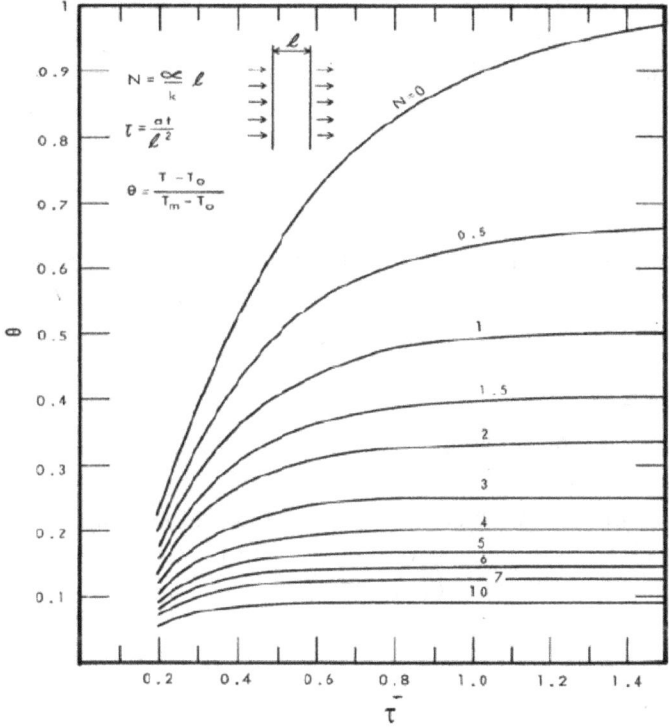

FIGURE 4.23. Graphical Heat Transfer Solution for ASTM E119 Exposure of Slab (Lie 1972)

81

Legend for Figure 4.23:
- a =   Thermal diffusivity of concrete (m^2/s)
- l =   Thickness of slab (m)
- N =   Biot number
- t =   Duration of exposure (s)
- T =   Temperature of concrete at position x (°C)
- T_m =   Mean exposure temperature (°C)
- T_o =   Initial temperature (°C)
- α =   Heat transfer coefficient ($W/m^2 \cdot °C$)
- τ =   Fourier number
- θ =   Dimensionless temperature

Limitations of the graphical analyses include the following:

- The graphical methods apply only for exposures associated with ASTM E119. Consequently, if this analysis is being conducted to assess the response of a concrete assembly to an alternative fire exposure, an approach described in Section 4.3.2 should be applied.
- Graphical solutions cannot be readily incorporated into computer-based methods that require the temperature of the assembly as input. So, while a graphical method may be relatively easy to apply, it may be tedious, and perhaps not very accurate (depending on the legibility of the graph), to input the results of the graphical analysis into a structural calculation.

Numerous algorithms are available to evaluate the thermal response of concrete slabs. Because the principal application of these algorithms is to assess the heat transfer in concrete sections exposed to non-standard fires, these methods are discussed in the next section.

4.3.2 Other Fire Exposures

A limited number of graphical solutions are available to describe the exposure from natural fires, e.g., the graphs presented by Pettersson, Magnusson, and Thor (1976) (see Section 2.5.3). Other solutions could potentially be developed, perhaps expanding the approach by Pettersson, Magnusson, and Thor.

Until other methods are developed, computer-based calculations are necessary to determine the temperature rise within concrete sections. Lie and Allen (1972) formulated a finite difference model to analyze the heating of circular reinforced concrete columns exposed to the standard fire. In addition, finite difference models were developed for concrete floor slabs (Lie 1978) and square reinforced concrete columns (Lie et al. 1984).

Ellingwood and Shaver (1980) and Ellingwood (1991) discuss the results of algorithms assessing the temperature distribution in concrete slabs and beams exposed to a variety of fire conditions.

Ahmed and Hurst (1995, 1998) applied a one-dimensional, finite difference analysis of the coupled heat and mass transfer through carbonate and siliceous aggregate concrete slabs and multi-layered gypsum wallboard-and-stud assemblies. Dehydration and evaporation phenomena and changes in porosity were considered by the model.

Heat transfer models that are widely available and have been applied to analyze the thermal response of assemblies exposed to fires include HEATING7 (Childs 1999), FIRES-T3 (Bresler Iding, and Nizamuddin 1977), TASEF-2 (Sterner and Wickstrom 1990) and HEAT (Munukutla 1989).

The numerical models are capable of assessing the heat transfer through concrete slabs as a result of exposure to the standard fire described in ASTM E119 or other defined fire exposures. Boundary conditions to characterize the fire are stipulated by convective and radiative parameters.

The numerical heat transfer models need to possess the following characteristics in order to address heating due to any fire exposure:

- Time-varying exposure conditions, including a wide range of possible conditions, ranging from a steady heat flux to non-linear conditions
- Temperature-dependent material properties

More advanced analyses should consider models such as those by Ahmed and Hurst (1995, 1998) that include mass and moisture transport and pore pressure analysis.

Limitations of the numerical analyses include the following:

- Most of the models assume that the material is homogeneous, with changes permitted only to account for variation in properties due to temperature. As such, changes due to the formation of cracks or pockets of moisture/water vapor are neglected. These can significantly affect the heat transfer process.
- Some thermal properties that vary with temperature may not be well characterized due to lack of data. If the properties are not known, a test program will need to be conducted to evaluate the unknown properties.
- Moisture migration and pressure build-up in pores are not addressed, except in specialized models. These effects will affect the heat transfer within a concrete assembly and will need to be accounted for either explicitly or implicitly.
- Spalling is not considered. Changes in the thickness of a concrete element due to spalling, formation of cracks, and changes in the surface characteristics can have a significant effect on heat transfer.

4.4 STRUCTURAL ANALYSIS

This section describes the analytical procedures that can be applied to all concrete construction to evaluate their structural response at elevated temperatures. Some differences exist between cast-in-place and precast construction, and between reinforced and prestressed concrete. These differences are noted in the appropriate locations in the section.

Evaluating the structural performance of concrete structures at elevated temperatures involves many of the usual propositions of the mechanics of reinforced and prestressed concrete (plane section behavior, strain compatibility between steel and concrete, etc.). The analytical procedures described here differ from the usual design calculations in that the properties of the materials (steel and concrete) at elevated temperatures are used in the calculations, and the effects of temperature on the distribution of forces in the structure must also be considered.

As an alternative to the use of calculation models described in these guidelines, a design may be based on the results of tests. The tests may be conducted in accordance with standard procedures, e.g., ASTM E119, or may be uniquely defined. Where unique tests are conducted, such tests need to be well documented, providing at least the same amount of information concerning the performance of the assembly as provided by calculations such as those described in this section.

4.4.1 Performance Criteria

Where load-bearing ability is required in the case of fire, concrete structures need to be designed and constructed to maintain their load-bearing function during the time period of interest. A structural analysis may be applied to assess load capacity or structural stability of fire-exposed sections. The performance criterion is stipulated as:

$$S_{n\theta} \geq Q_d \tag{4.4}$$

Where:

$S_{n\theta}$ = Nominal required strength at elevated temperature θ
Q_d = Demand

This statement of required performance is general and applies to all structural actions. The nominal strength of interest may be a bending moment, axial force, shear force, or torque. Similarly, the demand is represented by the internal bending moment, axial force, shear force, and torque at a section resulting from the self weight of the structure, superposed dead load, superposed live load, and the moment created by thermal effects. As a specific example, the nominal strength may be the nominal bending moment capacity and the demand may be the internal bending moment in equilibrium with external loads. In equation form, this strength requirement would be $M_{n\theta} \geq M_d$.

As an alternative to the strength criterion in Equation 4.4, deflection criteria may be adopted to assess serviceability of the concrete member. While deflection criteria are not explicitly stipulated in the standard test protocols used in North America to assess fire resistance, they may be adopted by the stakeholders of a performance-based analysis. Similarly, post-fire service-ability could be adopted by some stakeholders, though that is also not considered in current standard fire resistance test protocols.

4.4.2 Analysis of Flexural Members (Beams and Slabs)

A rational design method can be applied to calculate the strength of a flexural concrete element subjected to fire loading. Because the method of support is an important factor affecting structural behavior of flexural elements during a fire, the discussion that follows deals with three conditions of support:

1. Simply supported, thermally unrestrained, flexural members
2. Continuous flexural members
3. Simply supported, thermally restrained, flexural members

4.4.2.1 Simply Supported, Thermally Unrestrained, Flexural Members

The moment capacity analysis formulated for reinforced and prestressed concrete flexural members accounts for the composite nature of the assembly. The analysis of concrete slabs and beams is based on methods of analysis used in room temperature concrete design adapted for fire resistance analyses, with support from large-scale test data from the Portland Cement Association.

Figure 4.24 illustrates the behavior of a simply supported prestressed concrete beam exposed to fire from beneath. The ends of the beam are free to rotate, and the beam is free to elongate (thermally unrestrained). The steel reinforcement consists of straight prestressing strands located near the bottom of the beam. With the underside of the beam exposed to fire, the bottom expands more than the top, and the resulting curvature causes the beam to deflect downward. Also, the strength of the reinforcement and concrete near the bottom of the beam will decrease as the temperature rises. When the strength of the reinforcement reduces to less than that required to support the beam and any superposed load, flexural failure will occur. In essence, the moment demand, M_d, remains practically constant during the fire exposure, but the resisting moment, or nominal moment capacity, $M_{n\theta}$ is reduced as the reinforcement and the concrete near the bottom lose strength.

85

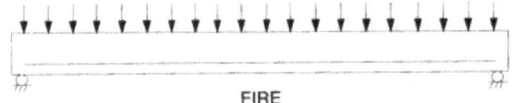

FIRE

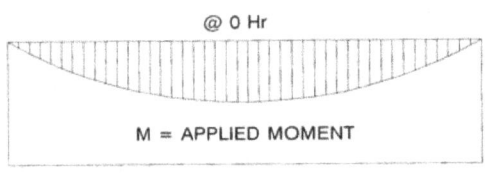

@ 0 Hr

M = APPLIED MOMENT

M$_n$ = MOMENT CAPACITY

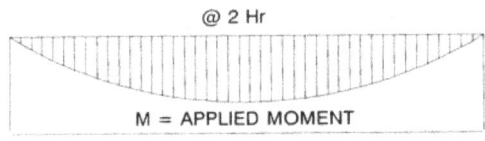

@ 2 Hr

M = APPLIED MOMENT

M$_{n\theta}$ = REDUCED MOMENT CAPACITY

FIGURE 4.24. Moment Diagram for Simply Supported, Thermally Unrestrained, Flexural Element (PCI 2004)

Because steel reinforcement is parallel to the axis of the beam, the design moment strength is constant throughout the length (CRSI 1980; PCI 1989; ACI 318 2005):

$$M_{n\theta} = A_s f_{s\theta}\left(d - \frac{a_\theta}{2}\right)$$ (4.5)

Where:

$M_{n\theta}$ = Moment capacity (lb-in.)
$f_{s\theta}$ = Critical stress in steel (psi)
A_s = Cross-sectional area of reinforcement steel (in.^{2})
d = Distance from extreme compression fiber to centroid of steel (in.)
a_θ = Depth of equivalent rectangular stress block (in.)

$$a_\theta = \frac{A_s f_{s\theta}}{0.85 f_c b}$$

f'_c = Compressive strength of concrete (psi)
b = Width of beam (in.)

in which θ signifies the effects of elevated temperatures. Note that A_s and d are not affected, but f_s is reduced based on the reinforcement temperature achieved (see Section 4.2). Similarly, the equivalent stress block depth, a, is reduced to maintain horizontal force equilibrium in the section as f_s is reduced. The concrete strength at the top of the beam, f'$_c$, is generally not reduced significantly because the temperature rise in the upper portion of the slab is typically relatively modest, if for no other reason than because of the heat transmission limit for fire resistance. For prestressed concrete, the tendon stress f_s can be determined from Equation 18-3 of the ACI 318-05 Code (ACI Committee 318 2005).

If the beam is uniformly loaded, the moment diagram is parabolic with a maximum value at mid-span of:

$$M = \frac{wl^2}{8}$$

(4.5)

Where:

        w   =   Dead plus live load per unit of length (k/in.)
        l =   Span length (in.)

As noted in Equation 4.4, flexural failure is assumed to occur when $M_{n\theta}$ is reduced to M_d. The usual ACI load factors and strength reduction factors are not applied (see Chapter 3). Thus, in this approach, the fire endurance depends on the applied loading and on the strength–temperature characteristics of the steel. In turn, the duration of the fire before the "critical" reinforcement temperature is reached depends on the protection afforded to the reinforcement.

Design problems involving the above equations can be solved utilizing data on the strength–temperature relationships for steel and concrete, and information on temperature distributions within concrete members during fire exposures (see the previous sections of this chapter).

The temperature of the reinforcement steel used in the analysis needs to be representative of the temperature of all positive reinforcing steel. Where reinforcement is located at multiple distances from an exposed surface, an "effective u," or $\bar{u}$, is used when applying Equation 4.5. The value of $\bar{u}$ is determined as:

$$\bar{u} = \frac{\sum\limits_{i=1}^{n} u_i A_i}{\sum\limits_{i=1}^{n} A_i}$$

(4.6)

For a beam, the representative temperature of the positive reinforcing steel is equal to the temperature attained in the concrete at an average position, $\bar{u}$, measured from the closest exposed surface. Where the difference in distance of a particular bar or strand to two exposed surfaces is not more than 25 %, for the purpose of this calculation, the distance to the surface for that bar or strand is ½ of the average of the two distances. This procedure does not apply to bundled bars or bundled strands.

The strength of the reinforcing steel is determined at the temperature resulting from the fire exposure, as determined following the heat transfer analysis methods described in the previous section and given the material properties presented in Section 4.2.3.

In addition, thin beams, such as in concrete single- or double-tee beams, are expected to experience a faster increase in temperature as a result of fire exposure. In some analysis methods, any portion of concrete that attains a threshold temperature is neglected, though the specific threshold temperature cited varies from 500 °C to 760 °C between publications and may also depend on the aggregate (Buchanan 2001; PCI 1989; CRSI 1980).

One important difference between cast-in-place and precast concrete construction is the amount of concrete cover required over the reinforcement required by ACI 318 (2005) or ASCE/SFPE 29 (1999). In general, precast concrete construction manufactured under plant control conditions requires less cover concrete than cast-in-place concrete. The justification for this difference is that better control can be exerted over formwork dimensions, placement of reinforcement, concrete quality, and curing procedures in a precast plant than in the field. While this better control may contribute to greater quality, a precast concrete element may have less concrete cover over the main longitudinal reinforcement (mild steel or prestressing steel), and therefore the steel temperature will increase more quickly than a similar cast-in-place concrete specimen. While the cast-in-place and precast construction may both satisfy minimum concrete cover requirements, there may be less reserve capacity in a precast element with less cover than a similar cast-in-place element.

4.4.2.2 Simply Supported, Thermally Restrained, Flexural Members

If a fire occurs beneath a portion of a large concrete floor (or roof), the heated portion of the floor will expand and push against the surrounding portion of the floor not involved in the fire. As a result, the unheated portion exerts compressive forces on the heated portion and creates restraint against thermal expansion. In general, restraint improves the performance of a member subjected to fire.

If restrained, the heated section will exert a force when attempting to expand against adjoining members. The force exerted on adjoining members is referred to as "thrust", T. In order to develop a moment from the thrust force and moment arm, d_T, two aspects must be true.

1. Adjoining members must be able to withstand the thrust force.
2. The line of action is not co-located with the neutral axis. If the line of action is below the neutral axis, the thrust acts similar to a prestressing force, thereby inducing a negative moment. A negative moment is beneficial to simply supported flexural members that need to resist an applied positive moment.

The nominal bending moment capacity of a thermally restrained beam or slab is:

$$M_{n,R} = M_{Th} + M_{n\theta} \tag{4.7}$$

$M_{n\theta}$ is determined by the equation presented above for simply supported members. The moment caused by the thrust resulting from the restraint, M_{Th}, is determined as:

$$M_{Th} = T_1\left(d_{Th} - \frac{a_{Th}}{2} - \Delta_1\right)$$ (4.8)

Where:

T_1 = Thermally-induced thrust
Δ_1 = Mid-span deflection
d_{Th} = Distance between the top of the member and the thrust line at the supports
a_{Th} = Depth of equivalent stress block for the thrust force

If the standard ASTM E119 fire exposure is considered, the thermally induced thrust, T_1, can be determined from the nomograms given in Figure 4.25 (PCI 1989):

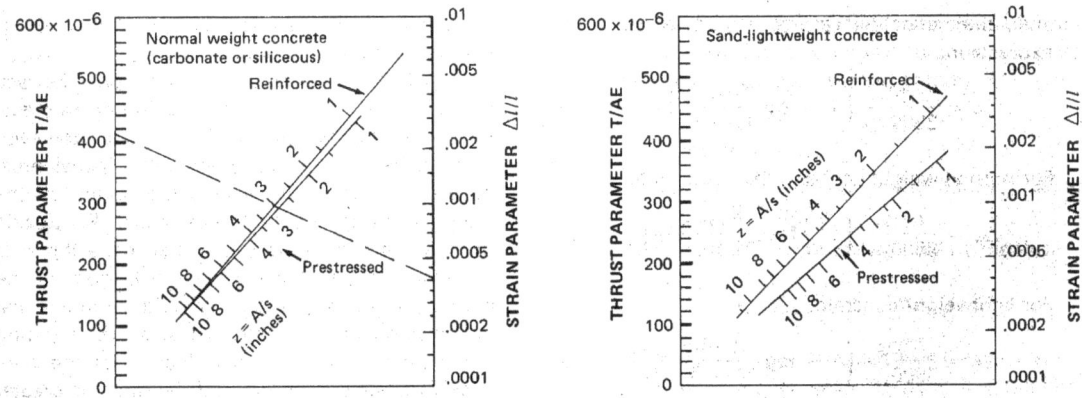

FIGURE 4.25. Nomograms Relating Thrust Parameter, Strain Parameter, and Ratio of Cross-Sectional Area to Heated Perimeter (PCI 1989)

Where:

T = Thrust force (lb)
A = Cross-sectional area (in^2)
E = Modulus of elasticity (psi)
Z = Dimensionless geometric parameter = A/S
A = Cross-sectional area (in^2)
S = Heated perimeter (in)

The distance between the top of the member and the line-of-action for the thrust force at the supports depends on the support conditions. Empirically, the position of the line of action of the thrust force was observed to be at or near the bottom of the stems in the multiple-T slabs. Thus, a conservative approach is to assume the location to be equal to 10 % of the depth of the slab above the bottom at the supports.

Where the standard ASTM E119 fire exposure is considered, the mid-span deflection, Δ_1, is determined as:

89

Normal-weight concrete:
$$\Delta_1 = \frac{l_1^2 \Delta_o}{3500 y_{b1}} \left(1 - 6\log\frac{T_1/A_1 E_1}{31} \right) \tag{4.9}$$

Lightweight concrete:
$$\Delta_1 = \frac{l_1^2 \Delta_o}{3500 y_{b1}} \left(1 - 6\log\frac{T_1/A_1 E_1}{74} \right) \tag{4.10}$$

Where:

Δ_1 = Deflection for a restrained slab (in.)

Δ_0 = Deflection for an unrestrained slab (in.) $= \dfrac{5wl^4}{384EI} = \dfrac{5MI^2}{48EZ_b y_b} = \dfrac{5f_{cb}l^2}{48Ey_b}$

y_{bl} = Distance from centroid to extreme fiber (in.)

If a fire exposure other than the standard fire exposure is considered, the thermally induced thrust, T_1, the line of action of the thrust force, and the mid-span deflection, Δ_1 need to be determined by an advanced computational method or test.

The depth of the equivalent compressive stress block for the thrust force, a_T, is determined as:

$$a_T = \frac{T + A_{ps} f_{ps}}{.85 f_c b} \tag{4.11}$$

d_T is estimated to be 90 % of the overall depth of the section, and Δ is determined from Figure 4.26. Both of these guidelines are only applicable given exposure conditions from the ASTM E119 standard test.

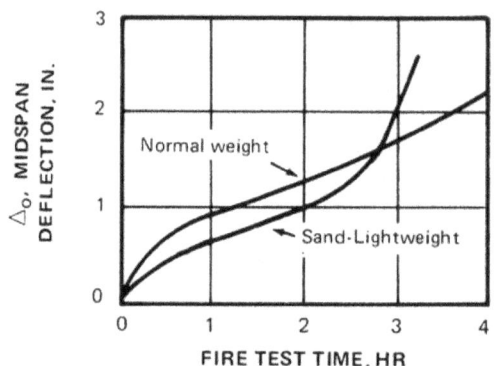

FIGURE 4.26. Estimate of Midspan Deflection, Reference Specimens (PCI 1989)

The actual location of the line of action the thrust force, d_T, for a concrete member depends on:

- Time duration of fire exposure
- Shape of the member
- Concrete compressive strength
- Amount of reinforcement
- Relative stiffness of the flexural member and the adjoining frame
- Amount of expansion permitted

4.4.2.3 Continuous Flexural Members

Continuous members undergo changes in reactions and internal forces when subjected to fire, referred to as "moment redistribution." Temperature gradients through the depth of the beam or slab create curvature, which, when integrated along the length of the element, creates deflection. The continuity of support condition creates reaction forces that are superposed with the reactions due to applied loads.

The moment diagrams for a two-span continuous beam whose underside is exposed to fire are presented in Figure 4.27. With the underside of the beam exposed to fire, the bottom of the beam expands more than the top. The resulting curvature increases the downward reaction at the interior support. This action results in a redistribution of moments: the negative moment at the interior support increases while the positive moments decrease.

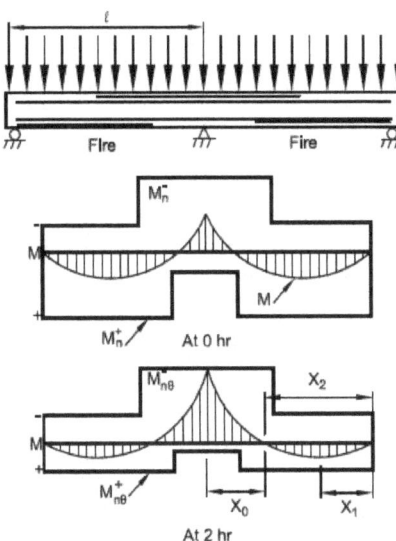

FIGURE 4.27. Moment Diagram for a Two-Span Continuous Beam (PCI 2004)

For the situation shown in Figure 4.27 (fire exposing the underside of the continuous beam or slab), the negative moment reinforcement remains cooler than the positive moment reinforcement because it is better protected from the fire. In addition, the redistribution that occurs during fire exposure may cause a sufficient increase in negative bending moment and yielding of the negative moment reinforcement. Thus, a relatively large increase in negative moment can

91

develop throughout the test. The resulting decrease in positive moment means that the positive moment reinforcement can be heated to a higher temperature before failure will occur. Therefore, the fire endurance of a continuous concrete beam is generally significantly longer than that of a simply supported beam having the same cover and the same applied loads.

It is possible to design the reinforcement in a continuous beam or slab for a particular fire endurance period. From Figure 4.27, the beam is expected to collapse when the positive moment capacity, $M^+_{n\theta}$, is reduced to the value of the maximum redistributed positive moment at a distance x_1 from the outer support.

The aspects to consider in a fire resistance analysis of a continuous slab are:

- Positive moment capacity
- Negative moment capacity
- Lengths of reinforcement

For a uniformly loaded, symmetrical interior span with equal end moments, the positive moment capacity is determined using the method outlined for simply supported sections. The required negative moment capacity is:

$$M^-_{n\theta} = \frac{wl^2}{8} - M^+_{n\theta} \tag{4.12}$$

The required length of the negative moment reinforcement is:

$$x_0 = \frac{l}{2} - \sqrt{\frac{2M^+_{n\theta}}{w}} \tag{4.13}$$

The available negative moment capacity is determined using the method outlined for simply supported sections with the following modifications. Because the concrete in compression will be located on the exposed side, the temperature of the concrete is determined at a location equal to half of the depth of the equivalent rectangular compressive stress block. The compressive strength of the concrete is then determined based on this temperature.

Figure 4.28 depicts a uniformly loaded beam or slab continuous (or fixed) at one support and simply supported at the other. Also shown is the redistributed applied moment diagram at failure.

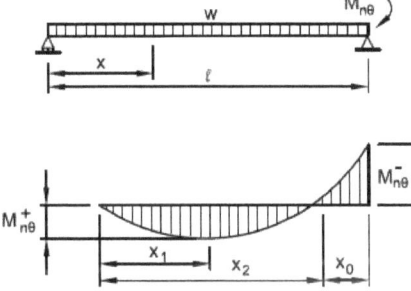

**FIGURE 4.28. Uniformly Loaded Member
Continuous (or Fixed) at One Support (PCI 2004)**

For a uniformly loaded end span, the positive moment capacity is determined as for an interior span, but the minimum, nominal negative moment capacity is determined as:

$$\mathrm{M}_{n\theta}^{-} = \frac{wl^{2}}{2} \pm wl^{2}\sqrt{\frac{2M_{n\theta}^{+}}{wl^{2}}} \qquad (4.14)$$

The required length of the negative moment reinforcement in the end span is determined as:

$$x_{0} = \frac{2M_{n\theta}^{-}}{wl} \qquad (4.15)$$

In most cases, redistribution of moments occurs early during the course of a fire, and the negative moment reinforcement can be expected to yield before the negative moment capacity has been reduced by the effects of fire. In such cases, the length of x_0 is increased, i.e., the inflection point moves toward the simple support. If the inflection point moves beyond the point where the bar stress cannot be developed in the negative moment reinforcement, sudden failure may result.

A symmetrical beam or slab in which the end moments are equal is illustrated in Figure 4.29.

93

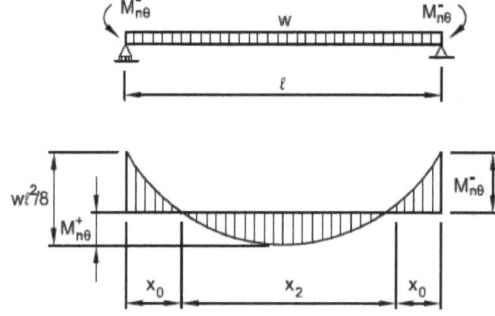

FIGURE 4.29. Uniformly Loaded Member Continuous at Supports (PCI 2004)

In this case,

$$\mathrm{M}_{n\theta}^{-} = \frac{wl^2}{8} - M_{n\theta}^{+} \qquad (4.16)$$

$$x_0 = \frac{l}{2} - \sqrt{\frac{8M_{n\theta}^{+}}{w}} \qquad (4.17)$$

To determine the maximum value of x_0, the value of w should be the minimum service load anticipated, and $\left(\dfrac{wl^2}{8} - \mathrm{M}_{n\theta}^{-}\right)$ should be substituted for $M_{n\theta}^{+}$.

For any given fire endurance period, the value of $M_{n\theta}^{+}$ can be determined by the procedures given above. Then the value of $\mathrm{M}_{n\theta}^{-}$ can be determined, followed by the necessary lengths of the negative moment reinforcement.

The amount of moment redistribution that can occur depends on the amount of negative moment reinforcement. Tests have clearly demonstrated that in most cases the negative moment reinforcement will yield, so the negative moment capacity is reached early during a fire test, regardless of the applied loading. The designer must exercise care to ensure that a secondary type of failure does not occur. To avoid a compression failure in the negative moment region, the amount of negative moment reinforcement should be small enough so that $\omega_\theta = \left(A_s f_{s\theta}\right)/b_\theta d_\theta f_{c\theta}'$ is less than 0.30, before and after reductions in f_y, b, d, and f_c' are taken into account. Furthermore, the negative moment bars or welded wire reinforcement must be long enough to accommodate the complete redistributed moment and change in the inflection points. It should be noted that the worst condition occurs when the applied loading is smallest, such as the dead load plus partial or no live load. It is recommended that at least 20% of the maximum negative moment reinforcement extend throughout the span.

94

Many of the precast concrete elements used in building structures (e.g., hollow-core planks, double-tee beams, and inverted tee beams) are used as simply supported elements. As such, the development of secondary shears due to thermal effects is often not a consideration. In contrast, much of the cast-in-place building construction is continuous, and consideration needs to be given to the effects of this continuity as described above.

4.4.3 Analysis of Compression Members

The axial-flexural behavior of concrete columns in fire may be governed by buckling strength. Allen and Lie (1974) evaluated the buckling of reinforced concrete columns by conducting an analysis of the strains through the cross section at mid-span. Stress resultants were applied to determine the strain distribution through the cross section. The total strain was related to the curvature in the column. Buckling behavior is noted when the radius of curvature decreases without limit. The analysis compared favorably with experimental data (Harmathy 1993).

4.4.3.1 Ultimate Load Calculation—Simplified Method

Dotreppe, Franssen, and Vanderzeypen (1999) evaluated the fire resistance of concrete columns based on tabulated data containing dimensions of the cross section and concrete cover. Dotreppe, Franssen, and Vanderzeypen developed a quick and efficient design method for determination of the ultimate load capacity in the case of a prescribed fire resistance, as well as determination of the fire resistance of a column in a building. This design formula is related to the plastic crushing load at elevated temperatures, which is reduced by the buckling coefficient for eccentrically loaded columns and a nonlinear amplification term for eccentric loads. The results are calibrated on experimental results. Spalling of concrete, slenderness ratio, and varying concrete cover are taken into account.

Limitations of this design method include the following (Dotreppe, Franssen, and Vanderzeypen 1999):

1. Columns with longitudinal bars with diameter $\phi \leq 25\text{mm}$.
2. Method is calibrated on experimental results.
3. $\beta_1(t)$ and $\beta_2(t)$ were considered using rectangular cross sections. However, the formulation can be extended to circular columns.
4. $N_{u\text{-theoretical}} / N_{u\text{-test}} \sim 0.4 - 1.4$; mean = 0.9.
5. $\beta_1(t)$ and $\beta_2(t)$ are limited to siliceous aggregate columns subjected to ISO 834, ASTM E119, and ULC S 101 fires.
6. λ (slenderness ratio) ≤ 100.
7. $0.04 \text{ m}^2 \leq A_c \leq 0.2 \text{ m}^2$.
8. $h/b \geq \frac{1}{2}$ (with $h \leq b$).
9. 20 mm $\leq c \leq 50$ mm.
10. $e \leq h/2$ (Note: Even if column is axially loaded, an eccentricity $e = 10$ mm should be adopted.)

The principal equation for this method is expressed as Equation 4.18:

$$N_u(t) = \gamma(t) * \eta(t) * N_p(t) \tag{4.18}$$

Where:

$\gamma(t)$ = Describes spalling as a function of time
$\eta(t)$ = Buckling coefficient for concentrically loaded columns
$N_p(t)$ = Plastic crushing load
$N_u(t)$ = Ultimate load capacity

For short columns: $\qquad\qquad N_u(t) = N_p(t) \tag{4.19}$

For long columns: $\qquad\qquad N_u(t) = \gamma(t) * \eta(t) * N_p(t) \tag{4.20}$

The following steps are used in the simplified method for ultimate load calculation:

Step 1: Calculate Plastic Crushing Load

$$N_p(t) = \beta_1(t) * A_c * f_c + \beta_2(t) * A_s * f_y \tag{4.21}$$

Determine correlation factors:

$$\beta_1(t) = \frac{1}{\sqrt{1 + (a_1 * t)^{a_2}}} \tag{4.22}$$

$$a_1 = 0.3 * A_c^{-0.5} \tag{4.23}$$

$$a_2 = A_c^{-0.25} \tag{4.24}$$

$$\beta_2(t) = 1 - \frac{0.9t}{0.046c + 0.111} > 0$$

Step 2: Calculate Spalling Factor

$\gamma(t) = 1 - 0.3t \qquad$ for t < 0.5 hour

$\gamma(t) = 0.85 \qquad$ for t > 0.5 hour

Step 3: Calculate Buckling Coefficient

Determine slenderness:

$$\text{for } \lambda \le 20 \qquad \chi(\lambda) = 1 - \frac{\lambda}{100} \tag{4.25}$$

for $20 < \lambda \le 70$ $\qquad$ $\chi(\lambda) = 0.80\left(\dfrac{20}{\lambda}\right)^{0.7\left(\frac{225-c}{200}\right)^{5}}$ $\qquad$ (4.26)

for $\lambda > 70$ $\qquad$ $\chi(\lambda) = 0.80\left(\dfrac{20}{\lambda}\right)^{0.7\left(\frac{\lambda}{70}\right)\left(\frac{225-c}{200}\right)^{5}}$ $\qquad$ (4.27)

Determine buckling coefficient: $\qquad$ $\eta(t) = \dfrac{\chi(\lambda)}{1 + \dfrac{10e/h}{\dfrac{1}{\chi(\lambda)} - 3.10^{-5}\lambda^{2}}}$ $\qquad$ (4.28)

where,

$\qquad$ e $\qquad$ = load eccentricity (mm)
$\qquad$ h $\qquad$ = smaller dimension of the cross section (mm)
$\qquad$ λ $\qquad$ = slenderness ratio
$\qquad$ c $\qquad$ = concrete cover (mm)
$\qquad$ Ac $\qquad$ = concrete area (m^2)
$\qquad$ As $\qquad$ = steel area (m^2)
$\qquad$ fc $\qquad$ = compressive strength of concrete (N/mm^2)
$\qquad$ fy $\qquad$ = yielding strength of steel (N/mm^2)
$\qquad$ t $\qquad$ = time or fire resistance (h)

Modifications to Simplified Formula:

1. $\qquad$ Aggregate factor described by Tan and Tang [2004] based on studies by Lie and Woollerton [1998] and Lie and Kodur [1996]:

$\qquad$ α_{agg} = 1.0 for siliceous; 0.9 for carbonate aggregate

2. $\qquad$ ASTM E119 fire exposure factor by Tan and Tang [2004]:

$\qquad$ α_{ISO} = 1.0 for ISO 834 fire; and 0.85 for ASTM E119 fire

Apply modifications to time (t) with t_e: $\quad$ $t_e = \alpha_{agg} * \alpha_{ISO} * t$

Example:

Ultimate load (N_u) calculation for a given fire resistance rating (R_f)

Given: A 200 mm x 300 mm column with 6 #4 bars (diameter of 12 mm) that is 3.90 meters in length. The column is eccentrically loaded, with hinges at both ends. The column has a fire resistance rating of 2 hours based on ISO 834 standard fire exposure

Parameters:

A_c = 200 x 300 mm   6 φ 12 mm   L = 3.90 m
c = 25 mm   e = 20 mm   hinged at both ends
f_c = 35.7 N/mm^2 f_y = 493 N/mm^2 R_f = 120 min

Calculation:

λ = kL/r = 67.5   Assume k = 1.0 (effective length)

A_c = 0.06m^2

A_s = 6.8*10^{-4} m^2

$a_1 = 0.3 * A_c^{-0.5} = 1.22$

$a_2 = A_c^{-0.25} = 2.02$

$$\beta_1 = \frac{1}{\sqrt{1 + (a_1 * t)^{a_2}}} = 0.38$$

$$\beta_2 = 1 - \frac{0.9t}{0.046c + 0.111} = -0.42$$

since $\beta_2 \geq 0$ $\beta_2 = 0$

Therefore,

$$20 < \lambda \leq 70 \qquad \chi = 0.80\left(\frac{20}{\lambda}\right)^{0.7\left(\frac{225-c}{200}\right)^5} = 0.34$$

98

Calculate buckling coefficient:

$$\eta = \cfrac{\chi(\lambda)}{1 + \cfrac{10 e/h}{\cfrac{1}{\chi(\lambda)} - 3.10^{-5} \lambda^2}} \qquad = \qquad 0.25$$

Calculate spalling factor:

$$\gamma(t) = 0.85 \qquad\qquad \text{Since } t > 0.5 \text{ hr}$$

Determine ultimate load:

$$N_u = \gamma * \eta * \left(\beta_1 * A_c * f_c + \beta_2 * A_s * f_y\right) = 173 kN$$

4.4.3.2 Beam–Column Calculation Method

Tan and Tang (2004) extend the Rankine method (typically used to describe the critical buckling load in steel column design) to predict the fire resistance of reinforced concrete columns subjected to fire conditions. This method does not include non-uniform temperature and stress distribution across member sections, and the induced mechanical stresses due to thermal expansion and thermal gradient. In addition, other secondary effects such as initial crookedness are ignored.

The general form of the Rankine model applied to reinforced concrete columns in fire conditions is as follows:

$$\frac{1}{P_{Rr}(t)} = \frac{1}{u_{pr} P_p(t)} + \frac{1}{u_{er} P_e(t)} \qquad\qquad (4.29)$$

where u_{pr} and u_{er} are the respective plastic and elastic critical load reduction factor and account for load eccentricity or applied moment (for axially loaded columns, u_{er} is typically unity). $P_p(t)$ and $P_e(t)$ are the respective plastic and elastic critical axial loads.

This method includes the calculation of the modified normalized slenderness ratio, $\Lambda_r(t)$. For reinforced concrete columns, this term is typically less than 0.5, which also explains the insignificance of secondary effects compared with steel columns.

$$\Lambda_r(t) = \sqrt{\frac{P_{pr}(t)}{P_{er}(t)}} = \sqrt{\frac{u_{pr}}{u_{er}}} * \Lambda(t) \qquad\qquad (4.30)$$

Where, $\Lambda(t)$ is the slenderness ratio.

The reinforced concrete column buckling curves can then be determined from the modified buckling coefficient, $N_r(t)$:

$$N_r(t) = \frac{P_R(t)}{P_{pr}(t)} = \frac{P}{P_{pr}(t)} = \frac{1}{1 + \Lambda_r(t)^2} \tag{4.31}$$

However, both the buckling coefficient and the normalized slenderness ratios depend on fire exposure time, in which the plastic collapse load $P_{pr}(t)$ and the elastic critical load $P_{er}(t)$ decrease with time.

To account for the reduction in plastic collapse load and elastic critical load, Tan and Tang introduce $P_{pr}(t)$ and $P_{er}(t)$ in terms of ambient loads $P_{pr}(0)$ and $P_{er}(0)$.

$$P_{pr}(t) = \phi_p(t) P_{pr}(0) \quad \text{and} \quad P_{er}(t) = \phi_e(t) P_{er}(0) \tag{4.32}$$

$\phi_p(t)$ and $\phi_e(t)$ are reduction factors accounting for the deterioration of material yield strength and elastic modulus. Modified normalized buckling and modified normalized slenderness ratio are also expressed in these terms, as follows:

$$N_r(0) = \frac{\phi_p(t)}{1 + \dfrac{\phi_p(t)}{\phi_e(t)} \Lambda_r^2(0)} \quad \text{and} \quad \Lambda_r(0) = \sqrt{\frac{P_{pr}(0)}{P_{er}(0)}} = \sqrt{\frac{u_{pr} P_p(0)}{u_{er} P_e(0)}} \sqrt{\frac{u_{pr}}{u_{er}}} * \Lambda(0) \tag{4.33}$$

Hertz (1993) suggests that $\phi_e(t) = \left[\phi_p(t)\right]^a$, where a is between 1 and 2. From Dotreppe, Franssen, and Vanderzeypen a=1. With a=1,

$$N_r(0) = \frac{\phi_p(t)}{1 + \Lambda_r^2(0)} \tag{4.34}$$

The plastic crushing load and elastic critical load can then be calculated, as follows:

$$P_p(0) = Q_c(0) + Q_{yr}(0) = 0.85 f'_c(0) A_c + f_{yr}(0) A_{sr} \tag{4.35}$$

$$P_e(0) = \frac{\pi^2 \left[0.2 E_c(0) I_c + E_{sr} I_{sr}\right]}{L_e^2} \quad \text{(ACI Committee 318 2005)} \tag{4.36}$$

Where:

$Q_c(0)$ and $Q_{yr}(0)$	=	Plastic crushing load from concrete and the steel reinforcement
f'_c and $E_c(0)$	=	Concrete strength and elastic modulus
$f_{yr}(0)$ and $E_{sr}(0)$	=	Yield strength and elastic modulus of steel reinforcement
A_c and A_{sr}	=	Area of concrete and steel rebar
I_c and I_{sr}	=	Second moment of area of concrete and steel rebar at centroid

Temperature effects on steel reinforcement and concrete materials are considered by applying the $\beta_1(t)$ and $\beta_2(t)$ from Dotreppe, Franssen, and Vanderzeypen (1999) to the material models. The plastic reduction factor, $\phi_p(t)$, is determined from Equations 4.29, 4.30, 4.32, and 4.33 and then used in Equation 4.31 to determine the normalized buckling coefficient. Then, the ultimate load can be calculated from the Rankine load formula.

Using a similar approach, the Rankine model is further applied to eccentrically loaded reinforced concrete columns in fire conditions. A bilinear P-M interaction diagram is constructed from the ultimate axial compressive load, the balance point, and the ultimate bending capacity. *Note:* The P-M interaction curve shrinks with increasing temperature. Temperature effects are accounted for by applying the $\beta_1(t)$ and $\beta_2(t)$ factors mentioned earlier.

Better correlations are achieved with this method than the Dotreppe, Franssen, and Vanderzeypen method (Tan and Tang 2004).

4.4.3.3 Mathematical Model

Lie and Celikkol (1991) evaluated the buckling of circular reinforced concrete columns. Lie and Irwin (1993) also evaluated the buckling of reinforced concrete columns with rectangular cross sections. In both studies, a mathematical model was proposed to predict the fire resistance of both circular and rectangular concrete columns. A thermal analysis was conducted using finite difference techniques to attain a temperature profile across the column cross section as a function of time. Effects of moisture were included for the temperature profiles. The strength of the column was determined by load deflection analysis assuming linear curvature from the pin end to the mid-height of the column, and a relationship between mid-height deflection and curvature. The axial strain was varied until the internal mid-height moment was in equilibrium with the applied moment. Load deflection curves were calculated for each time step during the fire exposure until the maximum load the column can carry was determined. Spalling was not considered in this study.

Alternatively, thermo-mechanical models, similar to that reported by Dotreppe, Franssen, and Vanderzeypen may be applied to assess the structural response of fire-exposed concrete columns.

The anticipated behavior of cast-in-place and precast concrete columns is similar. Precast columns are often reinforced with mild steel reinforcement, the same as for cast-in-place columns. In addition, a precast column may include a prestressing force to control stresses during handling and erection. Often this prestressing force is provided with debonded post-tensioned steel strands. The magnitude of the prestress is small and will not greatly influence behavior. Two principal differences in construction may create differences in behavior:

1. Concrete cover requirements are less for precast concrete columns than cast-in-place columns. Because the main longitudinal reinforcement is distributed around the perimeter of a column, all the reinforcement in a precast concrete column can be expected to increase in temperature (and therefore reduce in strength) more quickly than the reinforcement in a similar cast-in-place column.
2. Precast concrete columns are most often constructed as prismatic multistory elements. In fact, the cross-sectional dimensions of a precast concrete column often remain

constant throughout the height of a building structure. In contrast, the cross section dimensions of a cast-in-place column are more likely to change every several floors. This may result in a greater overstrength in the upper story columns of a precast concrete building than with a cast-in-place concrete building. This greater over-strength will provide greater inherent fire resistance for the upper story column.

4.4.4 Shear and Torsion

Little information exists in the published literature about the shear and torsion behavior of concrete members in fires. Most of the structural fire tests that have been performed have focused on the flexural behavior of beams and slabs and on the axial behavior of columns.

In the beam flexural tests, the beams were carrying shear force as well as bending moment, so the shear strengths of the beams were also evaluated in the tests. Of the many structural fire tests performed on simply supported elements (unrestrained and restrained), no shear failures were reported. Only one shear failure has been reported in a test of a continuous beam. However, in that test, the shear reinforcement was under-designed for the service load in the absence of fire. As explained earlier, in the case of continuous beams, additional reaction forces develop as a result of thermal effect, and these reaction forces alter the distribution of internal shear. Therefore, careful attention needs to be paid to the shear demands in continuous flexural members in fires.

In the absence of experimental data created for a particular design application, the behavior of concrete members in shear and torsion subjected to a fire loading should be calculated using the procedures in ACI 318 (2005), allowing for the effects of temperature on material properties and on the distribution of internal shear and torsion demands. In particular, this means that the effects of elevated temperatures on the yield strength of transverse reinforcement should be included in the calculation of the nominal shear strength, V_n, and nominal torsion strength, T_n. The influence of thermal effects on the magnitude of the internal shear force may be particularly important.

4.4.5 Spalling and Spalling Mitigation

Spalling refers to the delaminating or breaking away of a surface layer of concrete due to exposure to elevated temperature. There are many variations of spalling. Two types of concrete spalling that are more critical in structural performance point of view are explosive spalling and corner spalling. Explosive spalling, which tends to occur very suddenly and violently early in the fire exposure process, is widely observed in laboratory tests of small material specimens and structural elements (Phan, 2005, 2007, 2008; Bailey, 2002; Kalifa et al., 2000; Hertz, 1992) as well as real structures in accidental fires (Ulm et al., 1997). Corner spalling, which is often observed along the corners of rectangular or square concrete columns or beams, is a more gradual and less violent process.

While corner spalling is primarily the result of unrestrained thermal expansion in the transverse direction of the beams or columns, explosive spalling effects a larger surface area and has been shown to be primarily pore pressure-driven. The complex mechanism of explosive spalling and factors contributing to its occurrences has been explained in details by Phan (2005, 2007, 2007). Briefly, exposure to elevated temperature triggers the transformation and mass transport of moisture in concrete that results in the buildup of internal pore pressure and potential for explosive spalling as free water residing in the concrete pores and chemically-bound water in the concrete matrix undergo transformation from liquid to a gaseous phase, expand in volume, and transport through the concrete with increasing concrete temperature. High strength concrete is found to be more susceptible to explosive spalling than normal strength concrete due to its low permeability and thus reduced ability to successfully mitigate the buildup of internal pressure.

Spalling is significant in terms of structural fire performance because of the consequent reduction in the concrete cover causing the reinforcing steel to be heated more quickly, leading to premature loss of overall structural capacity. A secondary effect is the reduction in the overall slab thickness causing the unexposed surface temperature to increase more quickly. The amount of thickness reduction and the area of the slab affected are usually not large enough to cause problems with heat transmission through the slab (without otherwise causing very a substantial temperature increase in the steel reinforcement). Spalling itself is not identified as one of the endpoint criteria in ASTM E119.

The following is important information on explosive spalling that is useful for structural fire design and modeling purposes:

- The temperature of concrete (at the depth where spalling is expected to occur) when explosive spalling occurs is between 220 °C and 280 °C.
- Spalling depth typically varies between 25 mm and 75 mm.

The susceptibility of spalling by higher strength concrete may be abated through material design, reinforcement detailing, curing conditions, and use of additives such as polypropylene fibers. Studies (Phan, 2005, 2007, 2008) have shown that the addition of at least 1.5 kg/m³ of polypropylene fibers significantly reduced the buildup of pore pressure and tendency for explosive spalling in high strength concrete. Use of 135° bent seismic hook for stirrups in columns has also been found to reduce the tendency for spalling in columns (Kodur, 2005; ACI 216.1-07, 2007). The likelihood of spalling appears to increase for:

- High incident heat flux in the early stages of fire development
- Siliceous aggregate concretes and high-strength concretes
- Concretes with high moisture content (≥ 2 % by mass)
- Stress level:
  - Thin elements under high stress (i.e., prestressed concrete)
  - Abrupt changes in geometry and at connections to other structural members
- Concretes with low porosity/permeability

4.5 REFERENCES

Abrams, M.S. (1971), "Compressive Strength of Concrete at Temperatures to 1600F," *Temperature and Concrete*, SP-25, Detroit: American Concrete Institute, 33-58.

Abrams, M.S., and A.H. Gustaferro (1968), "Fire Endurance of Concrete Slabs as Influenced by Thickness, Aggregate Type and Moisture," Research Department Bulletin No. 223, Skokie, Ill.: Portland Cement Association.

Abrams, M.S., and A.H. Gustaferro (1969), "Fire Endurance of Two-Course Floors and Roofs," *J. of the American Concrete Institute* 66:2 (February 1969).

Abrams, M.S., A.H. Gustaferro, and E.A.B. Salse (1971), "Fire Tests of Concrete Joist Floors and Roofs," *RD Bulletin 006B*, Skokie, Ill.: Portland Cement Association.

ACI Committee 216 (1989), *Guide for Determining the Fire Endurance of Concrete Elements*, ACI 216R-89, Farmington Hills, Mich.: American Concrete Institute.

ACI Committee 216 (1997), *Determining Fire Resistance of Concrete and Masonry Construction Assemblies,* ACI 216.1-97, Farmington Hills, Mich.: American Concrete Institute.

ACI Committee 318 (1989), Building Code Requirements for Reinforced Concrete, ACI 318-89," *ACI Manual of Practice, 1990—Part 3: Use of Concrete in Buildings—Design Specifications, and Related Topics*, Detroit: American Concrete Institute.

ACI Committee 318 (2005), "Building Code Requirements for Reinforced Concrete," ACI 318-02, and Commentary, ACI 318R-05, Farmington Hills, Mich.: American Concrete Institute.

Ahmed, G.N., and J.P. Hurst (1995), "Modeling the Thermal Behavior of Concrete Slabs Subjected to the ASTM E119 Standard Fire Condition," *J. of Fire Protection Engineering* 7:4 (1995) 125-132.

Ahmed, G.N., and J.P. Hurst (1998), "Validation and Application of a Computer Model for Predicting the Thermal Response of Concrete Slabs Subjected to Fire," *ACI Structural J.* 95:5 (September-October 1998) 480-487.

Allen, D.E., and T.T. Lie (1974), "Further Studies of the Fire Resistance of Reinforced Concrete Columns," Technical Paper No. 416, NRCC 14047, Ottawa: National Research Council of Canada, Division of Building Research.

ASCE (1998), *Standard Calculation Methods for Structural Fire Protection*, ASCE/SFPE Std. 29-99, Reston, Va.: American Society of Civil Engineers.

ASTM (2000), "Standard Test Methods for Fire Tests of Building Construction and Materials," ASTM E119, West Conshohocken, Pa.: American Society for Testing and Materials.

Bennetts, I.D. (1981), "Elevated Temperature Behaviour of Concrete and Reinforcing Steel," Report No. MRL/PS23/81/001, Clayton, Victoria, Australia: BHP Melbourne Research Laboratories.

Bobrowksi, J., ed. (1978), *Design and Detailing of Concrete Structures for Fire Resistance*, Interim Guidance by a Joint Committee of the Institution of Structural Engineers and The Concrete Society, London: Institution of Structural Engineers.

Buchanan, A. (2001), *Structural Design for Fire Safety*, New York: John Wiley & Sons.

Childs, Kenneth (1999), Heating 7.3 User's Manual, Oak Ridge, Tenn.: Oak Ridge National Laboratory.

CRSI (1980), *Reinforced Concrete Fire Resistance*, Chicago: Concrete Reinforcing Steel Institute.

Cruz, C.R. (1966), "Elastic Properties of Concrete at High Temperatures," Research Department Bulletin 191, Skokie, Ill.: Portland Cement Association.

Dotreppe, Jean-Claude, Jean-Marc Franssen, and Yves Vanderzeypen (1999) "Calculation Method for Design of Reinforced Concrete Columns under Fire Conditions," *ACI Structural J.*, Technical Paper No. 96-2 (January-February 1999) 9-20.

EC2 (2002), "Design of Concrete Structures—Part 1.2, General Rules—Structural Fire Design," ENV 1992-1-2, CEN.

Ehm, H., and R. vanPostel (1967), "Tests of Continuous Reinforced Beams and Slab under Fire," *Proceedings, Symposium on Fire Resistance of Prestressed Concrete*, Translation available at S.L.A. Translation Center, Chicago: John Crerar Library.

Ellingwood, B. (1991), "Impact of Fire Exposure on Heat Transmission in Concrete Slabs," *J. of Structural Engineering* 117:6 (June 1991) 1870-1875.

Ellingwood, B., and J.R Shaver (1980), "Effects of Fire on Reinforced Concrete Members," *J. of the Structural Division*, Proceedings of the American Society of Civil Engineers, 106:ST11 (November 1980) 2151-2166.

Harmathy, T.Z. (1970a),"Thermal Properties of Concrete at Elevated Temperatures," *J. of Materials* 5:1 (March 1970) 47-74.

Harmathy, T.Z. (1970b), "Thermal Performance of Concrete Masonry Walls in Fire," *Fire Test Performance*, ASTM 464, West Conshohocken, Pa.: American Society of Testing and Materials, 209-243.

Harmathy, T.Z. (1993), *Fire Safety Design & Concrete*, New York: John Wiley & Sons.

Harmathy, T.Z., and L.W. Allen (1973), "Thermal Properties of Selected Masonry Unit Concretes," *ACI J.*, Proceedings V, 70:2 (February 1973) 95-104.

Hertz, K. (1993), "Simplified Calculation Method for Fire Exposed Concrete Structures," Supporting Document for CEN pr-ENV 1992-1-2, Technical University of Denmark.

Hosser, D., T. Dorn, and E. Richter (1994), "Evaluation of Simplified Calculation Methods for Structural Fire Design," *Fire Safety J.* 22 (1994) 249-304.

Huang, Zhaohui, and Andrew Platten, "Nonlinear Finite Element Analysis of Planar Reinforced Concrete Members Subjected to Fires," *ACI Structural J.*, Technical Paper No. 94-S25 (May-June 1997) 272-282.

Iding, R., Z. Nizamuddin, and B. Bresler (1977), FIre REsponse of Structures—Thermal 3-Dimensional Version, UCB FRB 77-15, Berkeley: University of California.

ISE (1978), "Design and Detailing of Concrete Structures for Fire Resistance," London: Institution of Structural Engineers.

Issen, Gustaferro, and Carlson (1970), "Fire Tests of Concrete Members: An Improved Method for Estimating Thermal Restraint Forces," ASTM STP 464, 153-179.

Kodur, V.K.R. (2005), Private Communication, January 31, 2005.

Kodur, V.K.R. (2005), "Guidelines for Fire Resistance Design of High-Strength Concrete Columns," *J. of Fire Protection Engineering* 15:2 (2005) 93-106.

Lie, T.T. (1972), *Fire and Buildings*, London: Applied Science Publishers, Ltd.

Lie, T.T. (1978), "Calculation of the Fire Resistance of Composite Concrete Floor and Roof Slabs," *Fire Technology* 14:1 (1978) 28-45.

Lie, T.T. (1992), *Structural Fire Protection*, Manual 78, Reston, Va.: American Society of Civil Engineers.

Lie, T.T., and D.E. Allen (1972), "Calculations of the Fire Resistance of Reinforced Concrete Columns," NRCC 12797, Ottawa: National Research Council of Canada, Division of Building Research.

Lie, T.T., and J.L. Woollerton (1988), "Fire Resistance of Reinforced concrete Columns: Test Results," Internal Report No. 569, Ottawa: National Research Council of Canada.

Lie, T.T., and B. Celikkol (1991), "Method to Calculate the Fire Resistance of Circular Reinforced Concrete Columns," *ACI Materials J.* 88:1 (January 1991) 84-91.

Lie, T.T., and R.J. Irwin (1993), "Method to Calculate the Fire Resistance of Reinforced Concrete Columns with Rectangular Cross Section," *ACI Structural J.*, Technical Paper No. 90-S7 (January-February 1993) 52–60.

Lie, T.T., and V.K.R. Kodur (1996), "Fire Resistance of Steel Columns Filled with Bar-reinforced Concrete," *J. of Structural Engineering* 122:1 (1996) 30-36.

Lie, T.T., T.D. Lin, D.E. Allen, and M.S. Abrams (1984), "Fire Resistance of Reinforced Concrete Columns," Technical Paper No. 378, Ottawa: National Research Council of Canada, Division of Building Research.

Milke, J.A. (1999), "Estimating Fire Performance of Concrete and Masonry Structural Members," *Proceedings of the 1999 Structures Congress*, Reston, Va.: ASCE, 377-380.

Munukutla, V.R. (1989), "Modeling of Thermal Performance of Concrete Walls," Christchurch, New Zealand: University of Canterbury, Department of Civil Engineering.

Ng, Ah Book, Saeed M. Mirza, and T.T. Lie (1990), "Response of Direct Models of Reinforced Concrete Columns," *ACI Structural J.*, Technical Paper No. 87-S32 (May–June 1990) 313-325.

PCA (1976), "Fire Endurance of Continuous Reinforced Concrete Beams," *RD Bulletin 072B*, Skokie, Ill.: Portland Cement Association.

PCI (2004), *PCI Design Handbook—Precast and Prestressed Concrete*, 6th ed., Chicago: Precast/Prestressed Concrete Institute, Industry Handbook Committee.

PCI (1989), *Design for Fire Resistance of Precast Prestressed Concrete*, 2nd ed., MNL-124-89, Chicago: Precast/Prestressed Concrete Institute.

Pettersson, O., S.E. Magnusson, and J. Thor (1976), "Fire Engineering Design of Structures," *Publication 50*, Swedish Institute of Steel Construction.

Phan, L.T. (1996), "Fire Performance of High-Strength Concrete: A Report of the State of the Art," NISTIR 5934, Gaithersburg, Md.: National Institute of Standards and Technology.

Phan, L., and N.J. Carino (2001), "Mechanical Properties of High-Strength Concrete at Elevated Temperatures, NISTIR 6726, Gaithersburg, Md.: National Institute of Standards and Technology.

Phan, L.T., and N.J. Carino (2003), "Code Provisions for High Strength Concrete Strength-Temperature Relationship at Elevated Temperatures," *RILEM Materials and Structures* 36 (March 2003) 91-98.

Phan, L.T. (2005), "Pore Pressure in High Strength Concrete at High Temperature," *Third International Conference on Construction Materials: Performance, Innovations and Structural Implications* (ConMat '05), Vancouver, Canada, August, 2005.

Phan, L.T., (2007) " Spalling and Mechanical Properties of High Strength Concrete at High Temperature," *Proceedings of the 5th International Conference on Concrete Under Severe Conditions: Environment and Loading* (CONSEC'07), Tours, France, June 2007, pp. 1595-1608, Volume 2.

Phan, L.T., (2008) "Pore Pressure and Explosive Spalling in Concrete," *RILEM Materials and Structures Journal* (Accepted for publication: January, 2008)

Sanjayan, G., and L.J. Stocks (1993), "Spalling of High-Strength Silica Fume Concrete in Fire," *ACI Structural J.*, Technical Paper No. 90-M18 (March-April 1993) 170-173.

Shirley, S.T., R.G. Burg, and A.E. Fiorato (1988), "Fire Endurance of High-Strength Concrete Slabs," *ACI Materials J.* (March-April 1988) 102-108.

Sterner, E., and U. Wickstrom (1990), "TASEF—Temperature Analysis of Structures Exposed to Fire," Fire Technology SP Report 1990: 05, Boras: Swedish National Testing Institute.

Tan, K.H., and C.Y. Tang (2004), "Interaction Formula for Reinforced Concrete Columns in Fire Conditions," *ACI Structural J.*, Technical Paper No. 101-S03 (January-February 2004) 19-28.

Underwriters' Laboratories of Canada (2004), *Fire Tests of Building Construction and Materials*, ULS Std. CAN4-S101, Toronto: Underwriters' Laboratories of Canada.

Weigler, H., and R. Fischer (1964), "Beton bei Temperaturen von 100 C bis 750 C," *Beton Herstellung und Verwedung*, Dusseldorf (February 2, 1964) 33-46.

4.6 Appendix A from Lie (1992)

Concrete Properties

Stress-strain relations

$$\text{for} \quad \varepsilon_c \le \varepsilon_{max} \qquad f_c = f_c' \left[1 - \left(\frac{\varepsilon_c - \varepsilon_{max}}{3 \cdot \varepsilon_{max}} \right)^2 \right]$$

$$\text{where} \qquad \varepsilon_{max} = 0.0025 + \left(6.0T + 0.04T^2 \right) \cdot 10^{-6}$$

$$\text{and} \qquad \text{for} \quad 0 \cdot C < T < 450 \cdot C \qquad f_c' = f_{co}'$$

$$\text{for} \quad 450 \cdot C \le T \le 874 \cdot C \qquad f_c' = f_c' \left[2.011 - 2.353 \left(\frac{T - 20}{1000} \right) \right]$$

$$\text{for} \quad T > 874 \cdot C \qquad f_c' = 0$$

Thermal capacity

$$\text{for} \quad 0 \le T \le 200 \cdot C \qquad \rho_c \cdot c_c = (0.005T + 1.7) \cdot 10^6 \cdot J \cdot m^{-3} \cdot C^{-1}$$

$$\text{for} \quad 200 \cdot C < T \le 400 \cdot C \qquad \rho_c \cdot c_c = 2.7 \cdot 10^6 \cdot J \cdot m^{-3} \cdot C^{-1}$$

$$\text{for} \quad 400 \cdot C < T \le 500 \cdot C \qquad \rho_c \cdot c_c = (0.013T - 2.5) \cdot 10^6 \cdot J \cdot m^{-3} \cdot C^{-1}$$

$$\text{for} \quad 500 \cdot C < T \le 600 \cdot C \qquad \rho_c \cdot c_c = (-0.013T + 10.5) \cdot 10^6 \cdot J \cdot m^{-3} \cdot C^{-1}$$

$$\text{for} \quad T > 600 \cdot C \qquad \rho_c \cdot c_c = 2.7 \cdot 10^6 \cdot J \cdot m^{-3} \cdot C^{-1}$$

Thermal conductivity

$$\text{for} \quad 0 \cdot C \le T \le 800 \cdot C \qquad k_c = -0.00085T + 1.9 \cdot W \cdot m^{-1} \cdot C^{-1}$$

$$T > 800 \cdot C \qquad k_c = 1.22 \cdot W \cdot m^{-1} \cdot C^{-1}$$

Coefficient of thermal expansion

$$\alpha_c = (0.008T + 6) \cdot 10^{-6}$$

Steel Properties

Stress-strain relations

for $\varepsilon_s \le \varepsilon_p$ $\qquad f_y = f \cdot \dfrac{(T, 0.001)}{0.001} \varepsilon_s$.

where $\qquad \varepsilon_p = 4 \cdot 10^{-6} \cdot f_{yo}$

and $\qquad f(T, 0.001) = (50 - 0.04T) \cdot \left[1 - \exp\left[(-30 + 0.03T) \cdot \sqrt{0.001} \right] \right] \cdot 6.9$

for $\varepsilon_s > \varepsilon_p$ $\qquad f_y = \dfrac{f(T, 0.001)}{0.001} \cdot \varepsilon_p + f\left[T, \left(\varepsilon_s - \varepsilon_p + 0.001 \right) \right] - f(T, 0.001)$

Thermal capacity

for $0 \le T \le 650 \cdot C$ $\qquad \rho_s \cdot c_s = (0.004T + 3.3) \cdot 10^6 \cdot J \cdot m^{-3} \cdot C^{-1}$

for $650 \cdot C < T \le 725 \cdot C$ $\qquad \rho_s \cdot c_s = (0.068T - 38.3) \cdot 10^6 \cdot J \cdot m^{-3} \cdot C^{-1}$

for $725 \cdot C < T \le 800 \cdot C$ $\qquad \rho_s \cdot c_s = (-0.086T + 73.35) \cdot 10^6 \cdot J \cdot m^{-3} \cdot C^{-1}$

for $T > 800 \cdot C$ $\qquad \rho_s \cdot c_s = 4.55 \cdot 10^6 \cdot J \cdot m^{-3} \cdot C^{-1}$

Thermal conductivity

for $0 \cdot C \le T \le 900 \cdot C$ $\qquad k_s = -0.022T + 48 \cdot W \cdot m^{-1} \cdot C^{-1}$

$T > 900 \cdot C$ $\qquad k_s = 28.2 \, W \cdot m^{-1} \cdot C^{-1}$

Coefficient of thermal expansion

for $T < 1000 \cdot C$ $\qquad \alpha_s = (0.004T + 12) \cdot 10^{-6} \cdot C^{-1}$

$T \ge 1000 \cdot C$ $\qquad \alpha_s = 16 \cdot 10^6 \cdot C^{-1}$

Water Properties
$\qquad \rho_w \cdot c_w = 4.2 \cdot 10^6 \cdot J \cdot m^{-3} \cdot C^{-1}$

Heat of vaporization:
$\qquad \lambda_w = 2.3 \cdot 10^6 \cdot J \cdot kg^{-1}$

Specifics of columns and furnace

$\varepsilon_f = 0.75$ emissivity of column furnace fire

$\varepsilon_c = 0.8$ emissivity of concrete

$KL = 2.0 \cdot m$ effective length of columns

$l = 3.5 \cdot m$ length of column that contributes to axial deformations

$\phi = 0.05$ concentration of moisture in insulation by volume

4.7 Commentary/Appendix

Information on T and d_T is based on experimental research by Issen, Gustaferro and Carlson (1970). The experimental program consisted of 40 standard fire resistance tests conducted by the Portland Cement Association.

The first 25 tests were conducted to provide a set of reference tests that could be used to obtain data to examine the accuracy of predictions from the analytical method. The 25 tests included 13 normal weight (carbonate) and 12 lightweight double-T slabs that were 16 feet long. The specimens were both prestressed and reinforced concrete designs. The expansion permitted in the tests ranged from 0.04 to 1.40 in. A diagram of a reference specimen is provided in Figure 4.30.

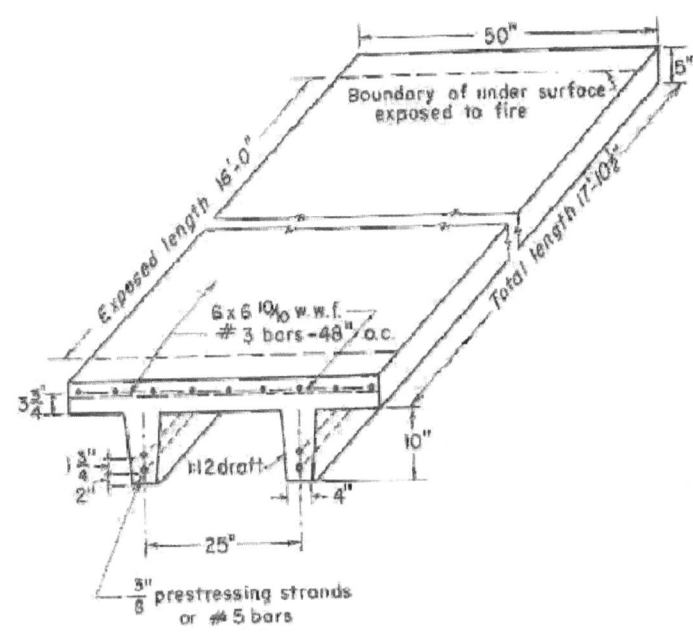

FIGURE 4.30. Reference Specimen (CRSI 1980)

111

The maximum thrust measured from the reference specimens is plotted in the graph in Figure 4.31. As expected, the thrust increased with a decrease in the amount of expansion permitted.

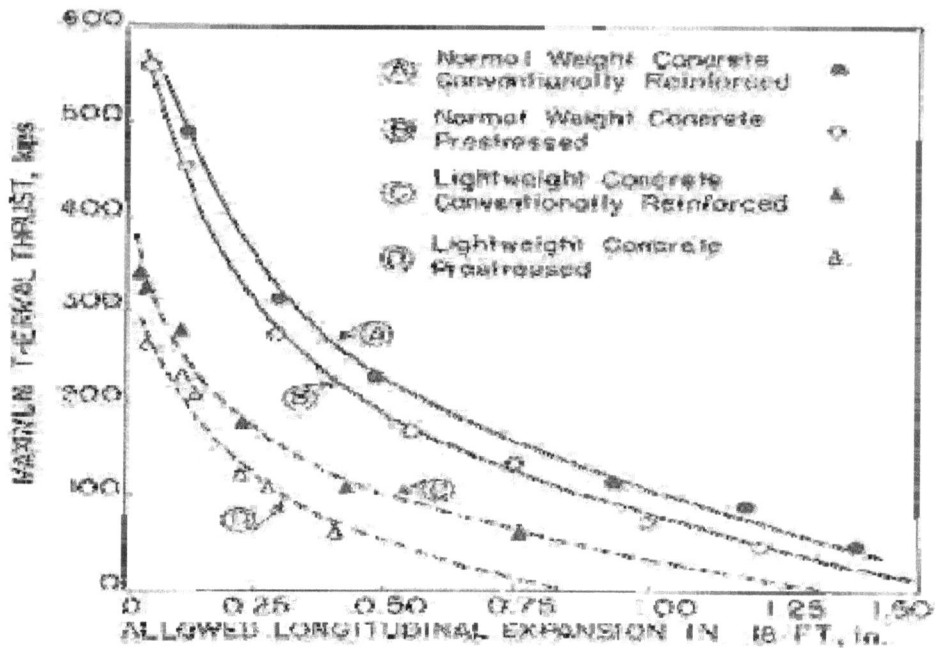

FIGURE 4.31. Maximum Thrust: Reference Specimens (CRSI 1980)

In the next phase of the experimental program, 15 tests were conducted with "correlation specimens." These specimens used different geometries and aggregates to observe differences in behavior. The analytical method developed from the reference specimens was adapted with the data from the correlation specimens for increased applicability.

The 15 specimens consisted of:

- 12 beams and slabs
  - Four 14 x 18 ft restrained slabs (restrained longitudinally and laterally)
  - 8 beams restrained longitudinally
- 3 miscellaneous specimens (siliceous)
  - 2 embedded electrical, underfloor ducts
  - 1 pan-joist with spray applied protection

The correlation specimens are depicted in Figure 4.32. Further details on the correlation specimens are:

- Properties of steel and concrete:
  - Prestressing steel: $f_{su} = 262$ ksi
  - Reinforcing steel: $f_y = 66$ ksi
  - Concrete: $f_c = 5$ ksi

112

- Applied load: factor of safety = 1.8, i.e., applied moment was 56% of ambient temperature moment capacity per American Concrete Institute code.

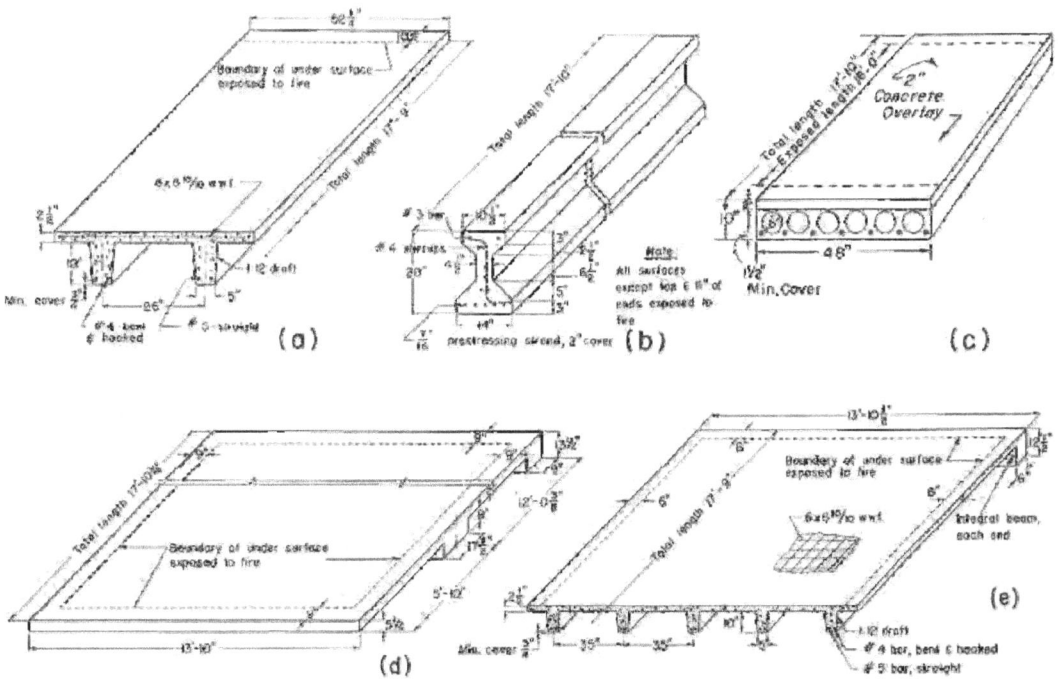

FIGURE 4.32. Correlation Specimens (CRSI 1980)

Agreement between the predicted and measured thrust is ± 15% for beams and –20 to +10 % for slabs, with a pan–joist slab at 80%.

4.8 Mathematical Model for Concrete Columns (Lie and Celikkol 1991)

Step 1: Calculation of Temperatures Using Finite Difference Method

Divide the column into concentric layers M (for circular column, Figure 4.33) and a triangular network (for rectangular sections, Figure 4.34). Assume the columns are exposed to the standard fire described by ASTM E119 or CAN4-S101 (ULC 2004).

113

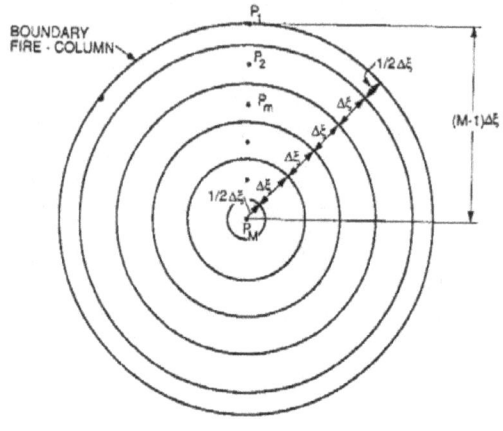

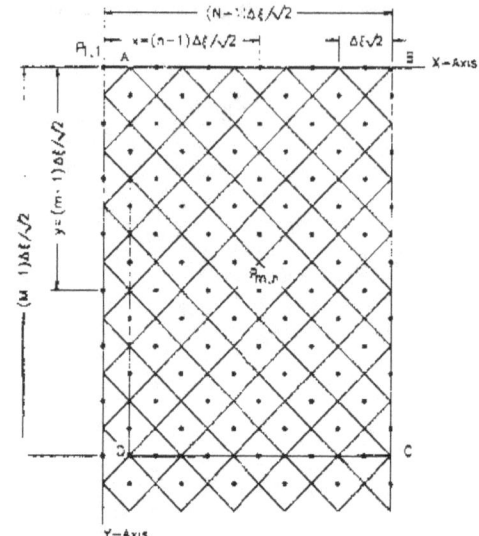

FIGURE 4.33. Arrangement of Layers in Circular Section

FIGURE 4.34. Arrangement of Elementary Layers in Rectangular Section

Equation for ASTM E119 time–temperature curve:

$$T_f^j = 20 + 750[1 - \exp(-3.79553\sqrt{t})] + 170.41\sqrt{t}$$

Temperature expression for outer concrete layer:

Circular section:

$$T_1^{j+1} = T_1^j + \frac{2(M-1)\Delta t}{(M-5/4)[(\rho_c c_c)_1^j + \rho_w c_w \phi_1^j]\Delta\xi}\{\sigma\varepsilon_f\varepsilon_c[(T_f^j + 273)^4 - (T_1^j + 273)^4]\}$$

$$- \frac{(M-3/2)(k_1^j + k_2^j)\Delta t}{M-5/4)[(\rho_c c_c)_1^j + \rho_w c_w \phi_1^j](\Delta\xi)^2}(T_1^j - T_2^j)$$

Rectangular section:

Elements along x-axis:

$$T_{1,n}^{j+1} = T_{1,n}^j + \frac{2\Delta t}{[(\rho_c c_c)_{1,n}^j + \rho_w c_w \phi_{1,n}^j]\Delta\xi^2}\left\{\left[\frac{(k_{2,(n-1)}^j + k_{1,n}^j)}{2}\right]\left[T_{2,(n-1)}^j - T_{1,n}^j\right]\right.$$

$$\left. + \left[\frac{(k_{2,(n+1)}^j + k_{1,n}^j)}{2}\right]\left[T_{2,(n+1)}^j - T_{1,n}^j\right] + \sqrt{2}\sigma\varepsilon_f\varepsilon_c\Delta\xi[(T_f^j + 273)^4 - (T_{1,n}^j + 273)^4]\right\}$$

114

Elements along y-axis:

$$T_{m,N}^{j+1} = T_{m,N}^{j} + \frac{2\Delta t}{[(\rho_c c_c)_{m,N}^{j} + \rho_w c_w \phi_{m,N}^{j}]\Delta \xi^2} \left\{ \left[\frac{(k_{(m-1),(N-1)}^{j} + k_{m,N}^{j})}{2} \right] \left[T_{(m-1),(N-1)}^{j} - T_{m,N}^{j} \right] \right.$$

$$\left. + \left[\frac{(k_{(m+1),(N-1)}^{j} + k_{m,N}^{j})}{2} \right] \left[T_{(m+1),(N-1)}^{j} - T_{m,N}^{j} \right] + \sqrt{2}\sigma\varepsilon_f\varepsilon_c\Delta\xi[(T_f^{j} + 273)^4 - (T_{m,N}^{j} + 273)^4] \right\}$$

Temperature expression for inside concrete layers:

Circular section:

$$T_m^{j+1} = T_m^{j} + \frac{\Delta t}{2(M-m)[(\rho_c c_c)_m^{j} + \rho_w c_w \phi_m^{j}]\Delta\xi^2} [(M - m + 1/2)(k_{m-1}^{j} + k_m^{j})(T_{m-1}^{j} - T_m^{j})$$

$$- (M - m - 1/2)(k_m^{j} + k_{m+1}^{j})(T_m^{j} - T_{m+1}^{j})]$$

Rectangular section:

$$T_{m,n}^{j+1} = T_{m,n}^{j} + \frac{\Delta t}{[(\rho_c c_c)_{m,n}^{j} + \rho_w c_w \phi_{m,n}^{j}]\Delta\xi^2} \left\{ \left[\frac{k_{(m-1),(n-1)}^{j} + k_{m,n}^{j}}{2} \right] \left[T_{(m-1),(n-1)}^{j} - T_{m,n}^{j} \right] \right.$$

$$+ \left[\frac{k_{(m+1),(n-1)}^{j} + k_{m,n}^{j}}{2} \right] \left[T_{(m+1),(n-1)}^{j} - T_{m,n}^{j} \right] + \left[\frac{k_{(m-1),(n+1)}^{j} + k_{m,n}^{j}}{2} \right] \left[T_{(m-1),(n+1)}^{j} - T_{m,n}^{j} \right]$$

$$+ \left[\frac{k_{(m+1),(n+1)}^{j} + k_{m,n}^{j}}{2} \right] \left[T_{(m+1),(n+1)}^{j} - T_{m,n}^{j} \right] \right\}$$

Temperature expression for center of concrete column:

Circular section:

$$T_M^{j+1} = T_M^{j} + \frac{2\Delta t}{[(\rho_c c_c)_M^{j} + \rho_w c_w \phi_M^{j}]\Delta\xi^2} (k_{M-1}^{j} + k_M^{j})(T_{M-1}^{j} - T_M^{j})$$

Temperatures along lines of symmetry:

$$T_{m,1}^{j+1} = T_{m,3}^{j+1} \qquad\qquad T_{(M+1),n}^{j+1} = T_{(M-1),n}^{j+1}$$

Rectangular section:

Calculate effect of moisture:

i)   Initial Volume of Moisture

Outer layer:

Circular section:

$$V_1 = \pi(M - 5/4)(\Delta\xi)^2 \phi_1$$

Rectangular (x-axis):

$$V_{1,n} = \frac{(\Delta\xi)^2}{2}\varphi_{1,n}$$

Rectangular (y-axis):

$$V_{m,N} = \frac{(\Delta\xi)^2}{2}\varphi_{m,N}$$

Inner layers:

Circular section:

$$V_m = 2\pi(M - m)(\Delta\xi)^2 \phi_m$$

Rectangular:

$$V_{m,n} = \frac{(\Delta\xi)^2}{2}\varphi_{m,n}$$

Center:

$$V_M = 1/4\pi(\Delta\xi)^2 \phi_M$$

ii)   Evaporation of volume with time

Outer Layer:

Circular section:

$$\Delta V_1 = \frac{2\pi\Delta t}{\rho_w \lambda_w}\left\{(M-1)\Delta\xi\sigma\varepsilon_f\varepsilon_c[(T_f^j + 273)^4 - (T_1^j + 273)^4]\right.$$
$$\left. - (M - 3/2)(\frac{k_1^j + k_2^j}{2})(T_1^j - T_2^j)\right\}$$

Rectangular (x-axis):

$$\Delta V_{1,n} = \frac{\Delta t}{\rho_w \lambda_w}\left\{\left[\frac{(k_{2,(n-1)}^j + k_{1,n}^j)}{2}\right]\left[T_{2,(n-1)}^j - T_{1,n}^j\right]\right.$$
$$+ \left[\frac{k_{2,(n+1)}^j + k_{1,n}^j}{2}\right]\left[T_{2,(n+1)}^j - T_{1,n}^j\right]$$
$$\left. + \sqrt{2}\sigma\varepsilon_f\varepsilon_c\Delta\xi[(T_f^j + 273)^4 - (T_{1,n}^j + 273)^4]\right\}$$

116

Rectangular (y-axis):

$$\Delta V_{N,m} = \frac{\Delta t}{\rho_w \lambda_w} \left\{ \left[\frac{(k^j_{(m-1),(N-1)} + k^j_{m,N})}{2} \right] \left[T^j_{(m-1),(N-1)} - T^j_{m,N} \right] \right.$$

$$+ \left[\frac{k^j_{(m+1),(N-1)} + k^j_{m,N}}{2} \right] \left[T^j_{(m+1),(N-1)} - T^j_{m,N} \right]$$

$$+ \left. \sqrt{2}\sigma \varepsilon_f \varepsilon_c \Delta \xi [(T^j_f + 273)^4 - (T^j_{m,N} + 273)^4] \right\}$$

Inner Layers:

Circular section:

$$\Delta V_m = \frac{2\pi \Delta t}{\rho_w \lambda_w} [(M - m + 1/2)(\frac{k^j_{m-1} + k^j_m}{2})(T^j_{m-1} - T^j_m)$$

$$- (M - m - 1/2)(\frac{k^j_m + k^j_{m+1}}{2})(T^j_m - T^j_{m+1})]$$

Rectangular section:

$$\Delta V_{m,n} = \frac{\Delta t}{\rho_w \lambda_w} \left\{ \left[\frac{k^j_{(m-1),(n-1)} + k^j_{m,n}}{2} \right] \left[T^j_{(m-1),(n-1)} - T^j_{m,N} \right] \right.$$

$$+ \left[\frac{k^j_{(m+1),(n-1)} + k^j_{m,n}}{2} \right] \left[T^j_{(m+1),(n-1)} - T^j_{m,N} \right]$$

$$+ \left[\frac{k^j_{(m-1),(n+1)} + k^j_{m,n}}{2} \right] \left[T^j_{(m-1),(n+1)} - T^j_{m,N} \right]$$

$$+ \left. \left[\frac{k^j_{(m+1),(n+1)} + k^j_{m,n}}{2} \right] \left[T^j_{(m+1),(n+1)} - T^j_{m,n} \right] \right\}$$

Center:

Circular section:
$$\Delta V_M = \frac{\pi \Delta t}{\rho_w \lambda_w} (\frac{k^j_{M-1} + k^j_M}{2})(T^j_{M-1} - T^j_M)$$

Step 2: Set Stability Criterion

The time increment must satisfy the following equation to minimize the error in temperature calculations:

$$\Delta t \leq \frac{(\rho_c c_c)_{min}(\Delta\xi)^2}{7200(k_{max} + h_{max}\Delta\xi)} \qquad h_{max} \approx 675 * W / m^2 C$$

Step 3: Calculation of Column Strength

Divide cross section into annular elements.

Determine temperature of annular concrete element:

Circular section (Figure 4.35): $\qquad (T_{m,n}^j)_{annular} = (\frac{T_m^j + T_{m+1}^j}{2})_{layer}$

Rectangular section (Figure 4.36): $\quad (T_{m,n})_{square} = \left[\frac{t_{(m+1),(n+1)}^j + T_{m,(n+2)}^j}{2}\right]_{triangular}$

Note: Temperature differences in steel rebar are often small. Steel rebar will be assumed to have the temperature of the concrete section at the location of the rebar center.

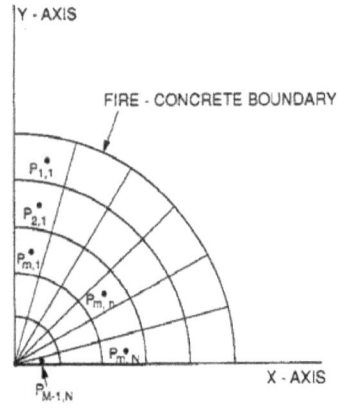

FIGURE 4.35. Arrangement in Quarter Sections of Circular Columns

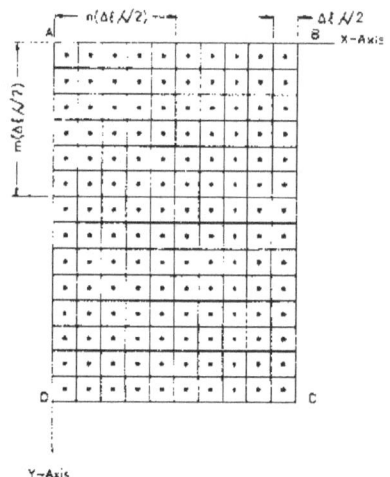

FIGURE 4.36. Stress–Strain Network in Quarter Section of Rectangular

Using the relationship between column mid-height deflection and curvature:

$$Y = \chi \frac{(KL)^2}{12}$$

118

For any given curvature, the axial strain is varied until the internal moment at the midsection is in equilibrium with the applied moment given by:

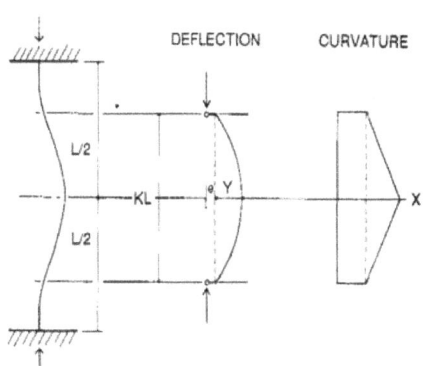

FIGURE 4.37. Load Deflection Analysis

Load x (deflection + eccentricity)

Assumptions:
1. Properties are as defined in Appendix.
2. Concrete has no tensile strength.
3. Plane sections remain plane.
4. The reduction of column length before exposure to fire, consisting of free shrinkage of concrete, creep, and shortening of column due to load, are negligible.
5. Spalling is not considered.

Equations for concrete:

Concrete strain to the right of y-axis:

$$(\varepsilon_c)_R = -(\varepsilon_T)_c + \varepsilon + \frac{x_c}{\rho}$$

Concrete strain to the left of y-axis:

$$(\varepsilon_c)_L = -(\varepsilon_T)_c + \varepsilon - \frac{x_c}{\rho}$$

Where:

$(\varepsilon_T)_c =$	Thermal expansion of concrete, mm-1
ε =	Axial strain of the column, mm-1
x_c =	Horizontal distance from center of the element to vertical plane through y-axis of column section, m
ρ =	Radius of curvature

Equations for steel:

Steel strain to the right of y-axis: $(\varepsilon_s)_R = -(\varepsilon_T)_s + \varepsilon + \frac{x_s}{\rho}$

Steel strain to the left of y-axis: $(\varepsilon_s)_L = -(\varepsilon_T)_s + \varepsilon + \frac{x_s}{\rho}$

119

Chapter 5

Design of Steel Structures

Farid Alfawakhiri, Ph.D., American Iron and Steel Institute

5.1 INTRODUCTION

Structural steel frames are widely used in virtually all types of buildings and industrial installations. Fire-resistant design of structural steel framing is often required, depending on the fire risks associated with the structure, the magnitude of potential losses due to structural failures caused by fire, and the accumulated performance record of similar structures in past fire incidents. While fire risks and consequences are discussed in earlier chapters, Beitel and Iwankiw (2002) provide a comprehensive review of fire-associated structural failure incidents in buildings higher than three stories.

5.1.1 Scope

This chapter reviews the background and practical methods of fire-resistant design in structural steel and consists of two parts. The first part covers the thermal and mechanical properties of structural steel at elevated temperatures. Additionally, several major groups of materials used for the fire protection of structural steel are discussed including insulating properties at elevated temperatures and practical guidance for their use. The second part discusses modern fire-resistant design methodologies for structural steel, covering methods based on standardized tests and the more sophisticated engineering analysis methods involving heat transfer and structural analysis at elevated temperatures.

This chapter does not include details for the fire protection design of open web steel joists (sometimes referred as "bar joists") or light-gauge cold formed steel structures. For these topics, the reader is referred to other publications (Design 2003, NAHB 2004).

5.2 MATERIAL PROPERTIES

The thermal and mechanical properties of most materials change substantially within the temperature range associated with building fires (up to 1200 °C). At elevated temperatures, structural steel undergoes physio-chemical processes that essentially change its composition, structure, and properties. Most of its material properties are temperature dependent and determination of the properties is sensitive to testing method and conditions such as heating rate, temperature gradient, strain rate, etc. An extensive summary of elevated temperature properties of steel and other building materials is provided in *The SFPE Handbook of Fire Protection Engineering* (Kodur and Harmathy 2002).

Harmathy (1983) was among the first researchers to systematically investigate and document the properties of common construction materials at elevated temperatures. He cited the lack of adequate knowledge of the behavior of building materials at elevated temperatures as the most disturbing trend in fire safety engineering, noting that there has been a tendency to use "national" ("typical," "proprietary," "empirical," etc.) values for material properties in numerical computations; in other words, values that ensure agreement between experimental and analytical results. Harmathy warned that this practice might lead to a proliferation of theories that lack general validity.

As more research has been conducted in the last two decades, the state of knowledge of the properties of common construction material at elevated temperature has improved considerably. However, applied materials research in the field of structural fire engineering continues to face numerous difficulties, especially in the area of materials used to protect steel. Some of the processes that affect these materials under heat exposure, such as moisture migration, intumescence, erosion, and spalling, are difficult to quantify or standardize in simple terms for engineering analysis. Many of the fire protection materials have proprietary formulations that are not readily available to the researcher or designer. The absence of generally accepted standard procedures to measure the heat transfer properties of fire protection materials (for the full range of temperatures associated with building fires) is a primary reason for the wide variation of conductivity and specific heat properties reported in the literature. However, standard procedures for regression analysis of the contribution of protection materials to fire resistance have been recently established in Europe (ENV 13381-1:2005, ENV 13381-2:2002, ENV 13381-4:2002).

In addition to the fact that elevated temperature properties of common fire protection materials are not generally available from the manufacturers, results of standard fire resistance tests of construction assemblies offer little technical information as well. This is because, in general, only the fire resistance rating achieved is reported to the public (the sponsor of the test, of course, has access to the entire test report). Thus, while it is possible to deduce at least bulk insulating properties of the fire protection materials from the rate of heating of the structural steel, without access to the actual recorded data, the tests provide little quantitative information.

5.2.1 Structural Steel

The performance of structural steel in fires is characterized by its thermal properties and mechanical properties. Thermal properties are necessary to predict the temperature rise in steel resulting from fire exposure and the resulting free thermal expansion, and include coefficients of thermal expansion, specific heat, and thermal conductivity. Prediction of mechanical behavior requires the stress-strain relationship of steel at elevated temperatures and may be represented by such parameters as elastic modulus, yield and ultimate strengths, and creep behavior. Poisson's ratio is generally considered to be invariant over the range of temperatures of interest and may be taken as 0.3. The effects on mechanical properties of the phase change that occurs generally between 700 °C and 850 °C are often neglected. For heat transfer calculations, steel density can be assumed constant at 7850 kg/m^3. The temperature-dependent thermal and mechanical properties of steel are discussed here.

5.2.1.1 Thermal Properties of Steel

5.2.1.1.1 *Thermal Expansion of Steel*

Steel expands when it is heated and the coefficient of thermal expansion (CTE) can be used to predict the expansion as a function of temperature, T. That is,

$$\Delta l/l = \alpha(T)\,\Delta T \qquad\qquad\qquad (5.1)$$

where,

Δl = *change in length due to temperature rise*
l = *initial length of the steel member*
$\alpha(T)$ = coefficient of thermal expansion
ΔT = change in temperature

The thermal elongation of a uniformly heated steel member, $\Delta l/l$, may be determined from the following, taken from Eurocode 3 (EN 1993-3-2):

$$\Delta l/l = -2.416(10^{-4})+1.2(10^{-5})T+0.4(10^{-8})T^2 \qquad \text{for } 20\,°C \leq T \leq 750\,°C \qquad (5.2a)$$
$$\Delta l/l = 11(10\text{-}3) \qquad \text{for } 750\,°C < T \leq 860\,°C \qquad (5.2b)$$
$$\Delta l/l = -6.2(10\text{-}3)+2.0(10\text{-}5)T \qquad \text{for } 860\,°C < T \leq 1200\,°C \qquad (5.2c)$$

where,

l = length of the steel member at 20 °C
Δl = change in length due to temperature rise
T = steel temperature, °C

The variation of the thermal elongation of steel with temperature is illustrated in Figure 5.1.

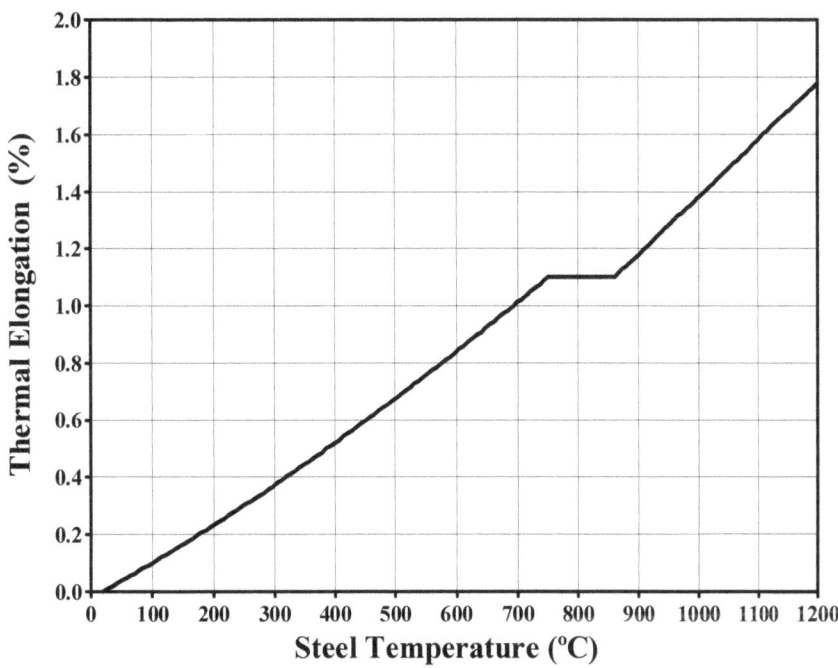

FIGURE 5.1. Thermal Elongation of Structural Steel at Elevated Temperatures (EN 1993-1-2: 2005)

The coefficient of thermal expansion, α, (see Eq. 5.1) varies with temperature and is equal to the slope of the relative elongation curve shown in Fig. 5.1. Note that steel begins to undergo a phase change at temperatures around 750 °C and, in fact begins to contract rather than expand. Because the phase change is not instantaneous, the behavior of steel in the range of 750 °C to 860 °C depends on the rate of heating. And, since the rate of heating is not known, the relative thermal elongation in this temperature range is taken to be constant and equal to 11 %. The coefficient of thermal expansion may be obtained by differentiating Eqs. 5.2a to 5.2c giving,

$$\alpha = 1.2(10^{-5}) + 0.8(10^{-8})T \qquad \text{for } 20\,°C \le T \le 750\,°C \qquad (5.3a)$$
$$\alpha = 0 \qquad \text{for } 750\,°C < T \le 860\,°C \qquad (5.3b)$$
$$\alpha = 2.0(10^{-5}) \qquad \text{for } 860\,°C < T \le 1200\,°C \qquad (5.3c)$$

Results for pure iron and for several low alloy steels have been reported by Touloukian et al. (1977). Banovic et al. (2005) note that the results are generally insensitive to composition and microstructure and recommend that the coefficient of thermal expansion be taken as that for pure iron which may be expressed in the form of a cubic polynomial as

$$\alpha\,(T) = \alpha_0 + \alpha_1 T + \alpha_2 T^2 + \alpha_3 T^3 \qquad (5.4)$$

where:
$$\alpha_0 = 7.3633 \times 10^{-6}$$
$$\alpha_1 = 1.8723 \times 10^{-8}$$
$$\alpha_2 = -9.8382 \times 10^{-12}$$
$$\alpha_3 = 1.6718 \times 10^{-16}$$

T is measured in Kelvin (°C+273) and the expression is valid over the range

$$300\,K < T < 900\,K \qquad \text{(approx. } 20\,°C < T < 630\,°C)$$

A plot of the instantaneous coefficient of thermal expansion (expressed in °C) for both Eq. 5.3a (i.e., $T \le 750\,°C$) and 5.4 is shown in Figure 5.2.

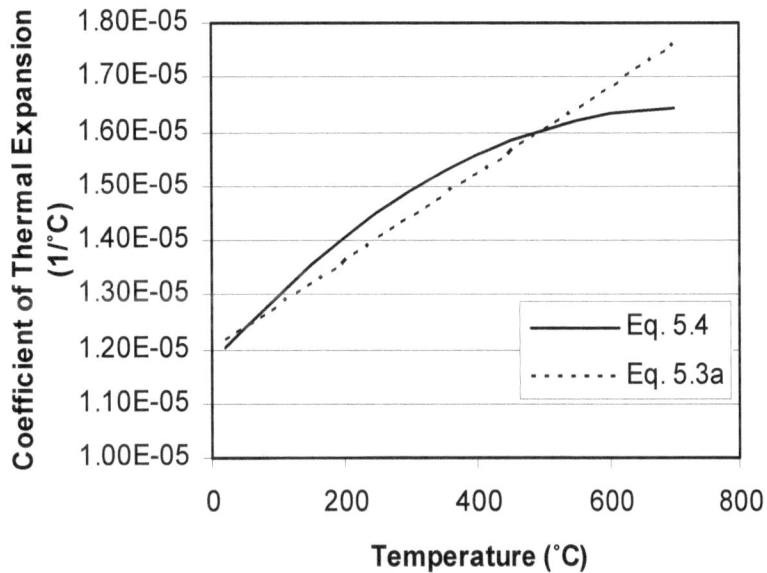

FIGURE 5.2. Instantaneous Coefficient of Thermal Expansion

Banovic recommends that, for temperatures in excess of 700 °C, the instantaneous coefficient of thermal expansion should be determined for the particular steel in question.

For simple calculation models, the relationship between thermal elongation and steel temperature may be considered to be linear (see Fig. 5.1) and the coefficient of thermal expansion (slope of the curve in Fig. 5.1) may be taken as a constant,

$$\alpha = 1.4(10^{-5}) / \,^\circ C \qquad\qquad (5.5)$$

5.2.1.1.2 Specific Heat of Steel

The specific heat of steel, c_s, valid for all structural grades, may be determined from the following, also taken from Eurocode 3 (EN 1993-3-2):

$$c_s = 425 + 7.73(10^{-1})T - 1.69(10^{-3})T^2 + 2.22(10^{-6})T^3 \text{ J/kg}^\circ C \quad \text{for } 20\,^\circ C \le T \le 600\,^\circ C \quad (5.6a)$$

$$c_s = 666 + \frac{13002}{738 - T} \text{ J/kg}^\circ C \qquad\qquad \text{for } 600\,^\circ C < T \le 735\,^\circ C \quad (5.6b)$$

$$c_s = 545 + \frac{17820}{T - 731} \text{ J/kg}^\circ C \qquad\qquad \text{for } 735\,^\circ C < T \le 900\,^\circ C \quad (5.6c)$$

$$c_s = 650 \text{ J/kg}^\circ C \qquad\qquad \text{for } 900\,^\circ C < T \le 1200\,^\circ C \quad (5.6d)$$

where,

      T =   steel temperature, °C

The variation of the specific heat with temperature is illustrated in Figure 5.3.

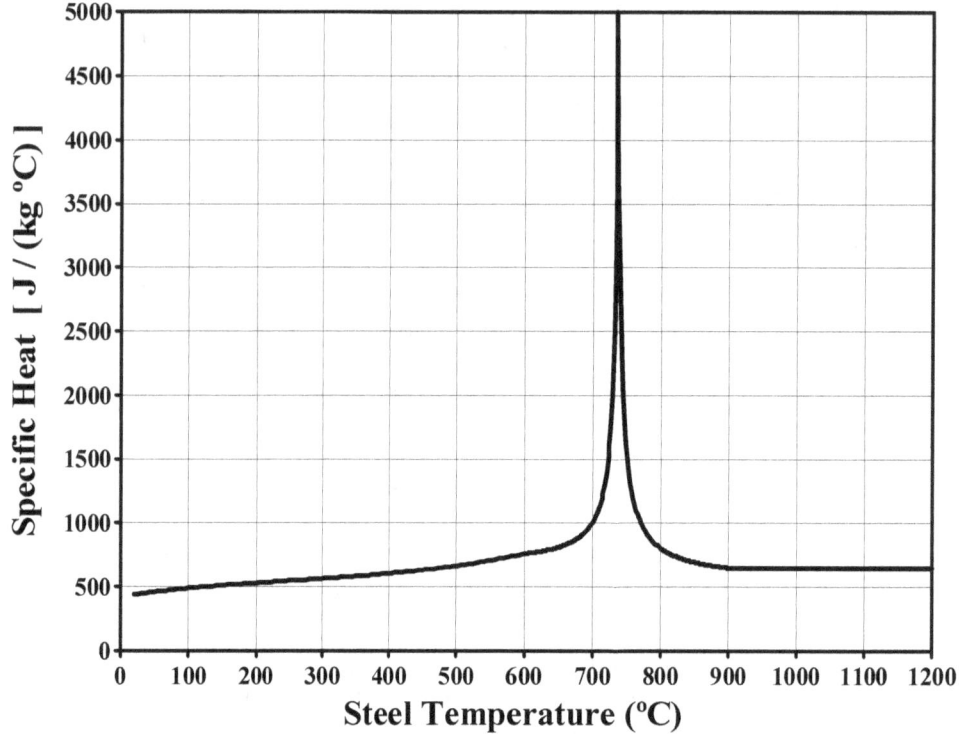

FIGURE 5.3. Specific Heat of Structural Steel at Elevated Temperatures (EN 1993-1-2: 2005)

In simple calculation models, the specific heat of steel may be considered to be independent of temperature with an average value of c_s = 600 J/kg°C.

5.2.1.1.3 Thermal Conductivity of Steel

Thermal conductivity properties of steel are required for determining the temperatures of steel members subjected to heat flux or fire exposure. Unlike the thermal properties just discussed, coefficient of thermal expansion and specific heat, thermal conductivity is affected by the microstructure of the steel. Fig. 5.4 shows the thermal conductivity as a function of temperature for twelve low-alloy steels as reported by Banovic (2005). This figure serves to illustrate the variability of thermal conductivity for various steel compositions and also plots the expression recommended in Eurocode 3 as described next.

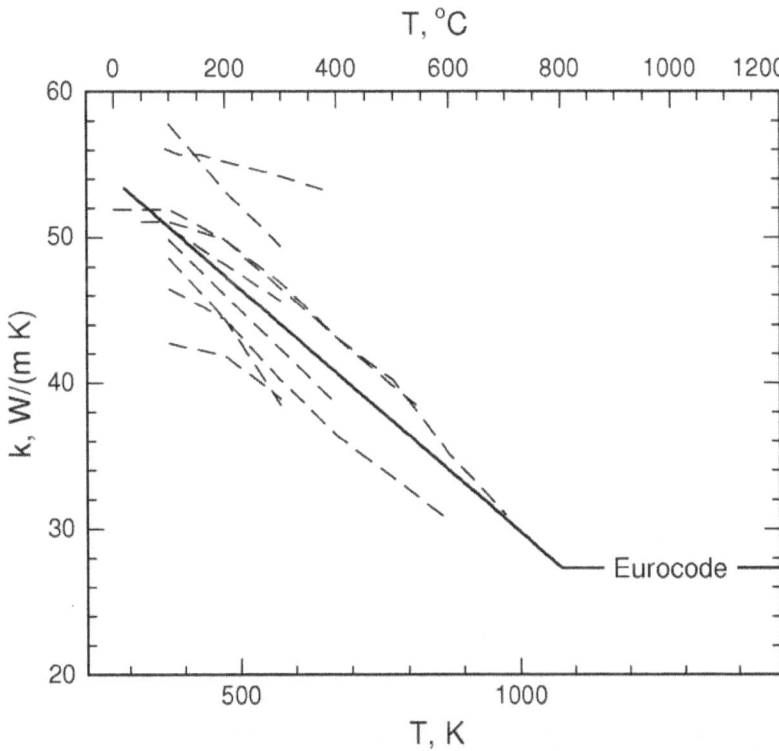

FIGURE 5.4. Thermal Conductivity of 12 Low-alloy Steels as a function of Temperatures (from Banovic 2005)

Eurocode 3 recommends that the thermal conductivity of steel, k_s, independent of structural grades, be determined from the following:

$$k_s = 54 - 3.33(10^{-2})T \quad W/m°C \qquad \text{for } 20\ °C \leq T \leq 800\ °C \qquad (5.7a)$$

$$k_s = 27.3 \quad W/m°C \qquad \text{for } 800\ °C < T \leq 1200\ °C \qquad (5.7b)$$

where,
$\quad$ T $\ =\ $ steel temperature, °C

The variation of the thermal conductivity with temperature, as recommended by Eurocode, is illustrated in Figure 5.5.

127

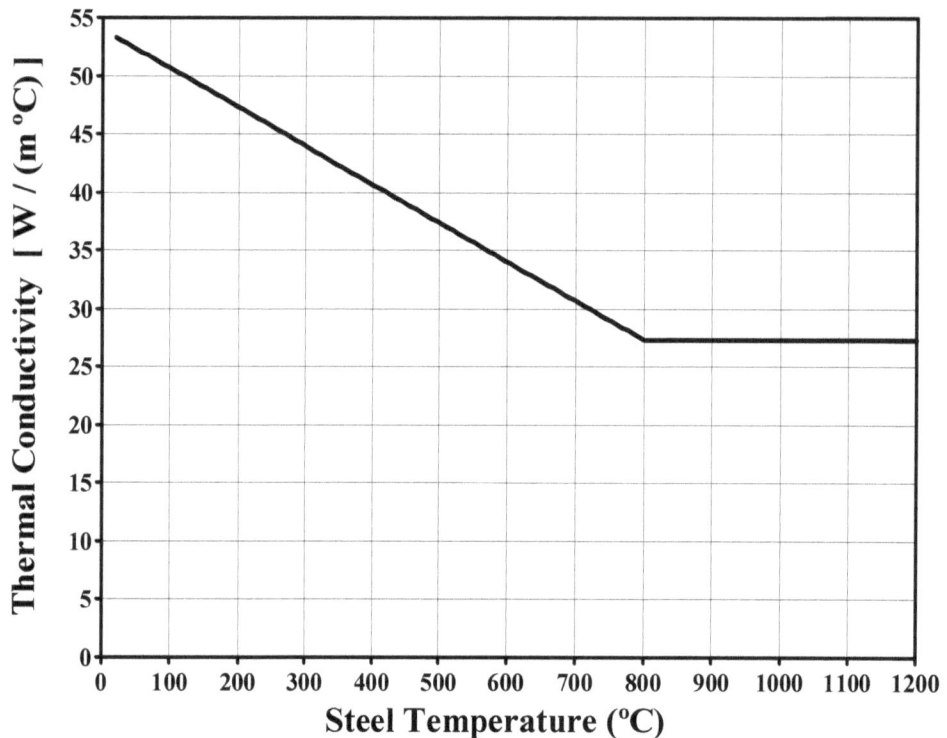

FIGURE 5.5. Thermal Conductivity of Structural Steel at Elevated Temperatures (EN 1993-1-2: 2005)

In simple calculations models, the thermal conductivity of steel may be considered to be independent of temperature with an average value of $k_s = 45$ W/m°C.

5.2.1.2 Mechanical Properties of Steel

Structural steel begins to lose its strength and stiffness at temperatures above 300 °C and eventually melts at about 1500 °C (Lawson and Newman 1990). The mechanical properties of steel at temperatures above 450 °C are strongly affected by creep, i.e., both stress and temperature histories influence steel deformations. In the combined heating and deformation of steel, the total strain in this temperature range can be separated into three components: the thermal strain, the instantaneous stress-related strain, and the time-dependent creep strain. Therefore, mechani-cal properties of steel can be established by following a number of different test procedures. The three main test parameters are the heating process, application and control of load, and control of strain. A clear distinction is made between isothermal "steady-state" tests and anisothermal "transient" tests (Anderberg 1983).

During the steady-state tests, the specimen is first heated to the intended steady temperature (the specimen is allowed to expand freely during this initial heating period) and then loaded using one of the following methods:
- Stress rate controlled test
- Strain rate controlled test

128

- Creep test, where the specimen is loaded to the intended level and kept constant while the deformations are recorded over time
- Relaxation test, where the specimen is loaded to the intended initial strain that is kept constant while the reduction in stress is recorded over time

In the United States, the first two types of steady-state tests mentioned above are standardized under ASTM designation E21-92. Steady-state test data for North American structural steel grades have been reported in the literature (Harmathy and Stanzak 1970, Brockenbrough and Johnston 1981).

During the transient tests, the specimen is first loaded to the intended level of stress which is then kept constant, while the specimen is steadily heated and the temperature–deformation relationship is recorded. The stress-strain relationship for a particular temperature can then be obtained by interpolation from a family of transient test curves for different stress levels. Another variation of transient tests is where the specimen is first restrained against thermal expansion (i.e., initially not loaded) and then steadily heated while the temperature–load relationship is recorded (compressive load develops in this test due to the restrained thermal expansion of the specimen).

During a building fire, the loaded steel structure is subjected to transient processes with varying temperature and stress. Therefore, transient tests could be claimed to be more realistic, and many researchers consider transient test data as an essential part of information for fire-resistant design purposes. Kirby and Preston (1988) demonstrated that, at small strains, transient data produce slightly more conservative results than steady-state tests. However, at larger strains there was negligible difference between the minimum properties derived from either type of test.

In the late 1970s, RILEM Committee 44-PHT over a five-year period conducted the first comprehensive international survey of high-temperature test data for structural and other steel grades, including North American grades. The final report (Anderberg 1983) concluded that the relative decrease in strength at elevated temperatures was almost the same for all structural grades, regardless of the initial strength of steel at room temperature. This observation led to the development of a single strength retention model for structural steel (of all grades) at elevated temperatures, first adopted by ECCS (1983). The model implicitly accounted for high-temperature creep associated with heating rates expected for steel during fires (2 °C/min to 50 °C/min). Later, more detailed experimental and analytical studies by Cooke (1988); Jerath, Cole, and Smith (1980): Kirby (1983); Kirby and Preston (1988) and others eventually led to the adoption of a complete stress–strain relationship model for structural steel at elevated temperatures in Eurocode 3 (EN 1993-1-2: 2005) and ECCS (2001).

As part of its investigation into the collapse of the World Trade Center towers (WTC 1 and 2) and Building 7 (WTC 7), NIST developed an elevated temperature true stress-true strain relationship based on data reported in the technical literature and on tests it conducted. Both the Eurocode and NIST approaches are described here.

5.2.1.2.1 Stress-Strain Relationship - Eurocode Approach

In the Eurocode 3 approach, three temperature-dependent parameters are used to define the stress-strain behaviour of steel at elevated temperatures:
1. The slope of the linear elastic range, E_T
2. The proportional limit stress, f_p
3. The yield stress, f_y

The yield stress is taken as the stress at 2 % strain ($\varepsilon = 0.02$) and is termed the "effective yield stress." An ellipse is fit between the proportional limit and the effective yield stress and beyond a strain of 2 %, the stress strain relationship is flat up to a strain of 15 % ($\varepsilon = 0.15$). The temperature-dependence of each parameter is normalized to a room temperature value; $E = 210,000$ MPa for E_T, and F_y at 20 °C determined from the $\varepsilon = 0.002$ (0.2 %) offset yield strength.

Accordingly, the stress f at strain ε for structural steel at elevated temperatures may be determined from the following expressions:

$$
\begin{array}{lll}
f = \varepsilon\, E_T & \text{for } 0 \leq \varepsilon < \varepsilon_p = f_p / E & (5.8a) \\
f = (b/a)\,(\, a^2 - (0.02 - \varepsilon)^2\,)^{0.5} + f_p - c & \text{for } \varepsilon_p \leq \varepsilon \leq 0.02 & (5.8b) \\
f = f_y & \text{for } 0.02 < \varepsilon \leq 0.15 & (5.8c) \\
f = 20\, f_y\,(0.2 - \varepsilon\,) & \text{for } 0.15 < \varepsilon \leq 0.20 & (5.8d) \\
f = 0 & \text{for } \varepsilon \geq 0.20 & (5.8e)
\end{array}
$$

where:

$$
\begin{array}{ll}
a^2 = (0.02 - \varepsilon_p)\,(0.02 - \varepsilon_p + c / E_T) & (5.8f) \\
b^2 = E_T\,(0.02 - \varepsilon_p)\, c + c^2 & (5.8g) \\
c = (f_y - f_p)^2 / (\, E_T\,(0.02 - \varepsilon_p) - 2\,(f_y - f_p)\,) & (5.8h)
\end{array}
$$

Table 5.1 gives, for steel elevated temperatures T, the reduction factors k_T to be applied to the appropriate values of the steel modulus of elasticity, E, and the specified minimum yield stress F_y at 20 °C in order to determine the parameters in Equations 5.8. For intermediate values of the temperature, linear interpolation may be used. For simple computation models, the proportional limit could be taken as the yield stress, i.e., $f_p = f_y$.

TABLE 5.1. Reduction Factors k_T for Stress–Strain Relationships of Structural Steel at Elevated Temperatures

Steel Temperature T (°C)	$k_{E,T} = E_T / E$	$k_{p,T} = f_p / F_y$	$k_{y,T} = f_y / F_y$	$k_{u,T} = f_u / F_y$
20	1.00	1.00	1.00	1.25
100	1.00	1.00	1.00	1.25
200	0.90	0.807	1.00	1.25
300	0.80	0.613	1.00	1.25
400	0.70	0.42	1.00	1.00
500	0.60	0.36	0.78	0.78
600	0.31	0.18	0.47	0.47
700	0.13	0.075	0.23	0.23
800	0.09	0.05	0.11	0.11
900	0.0675	0.0375	0.06	0.06
1000	0.045	0.025	0.04	0.04
1100	0.0225	0.0125	0.02	0.02
1200	0.00	0.00	0.00	0.00

Figure 5.6 illustrates the reduction factors listed in Table 5.1.

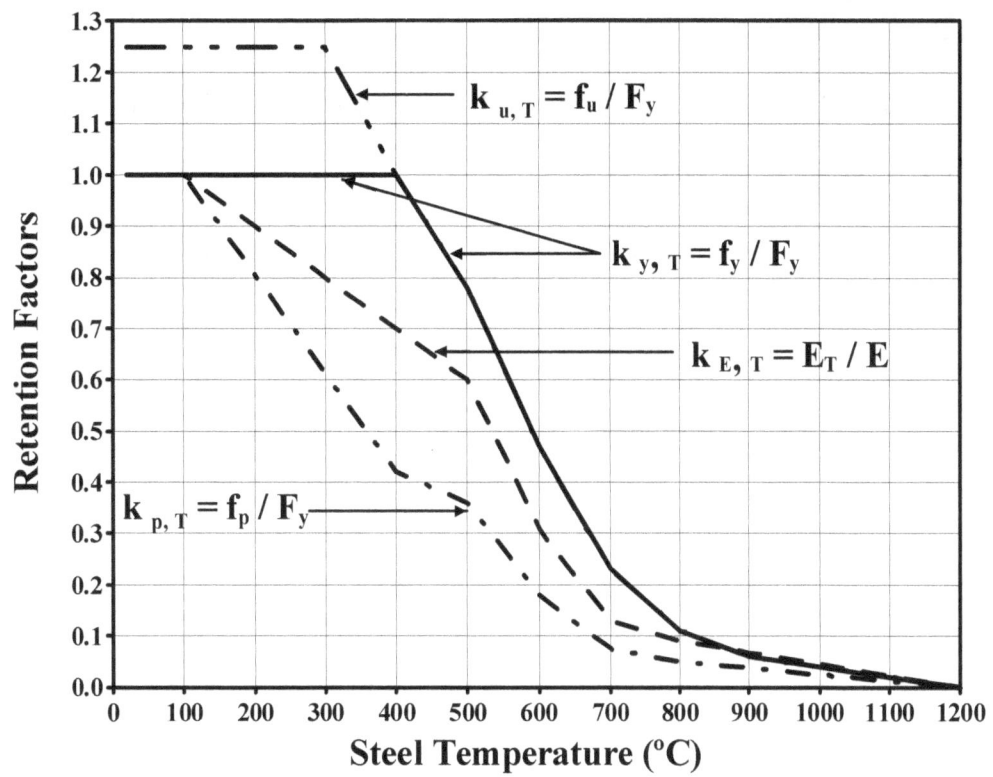

FIGURE 5.6. Strength and Elasticity Reduction Factors for Structural Steel at Elevated Temperatures (EN 1993-1-2: 2005)

For temperatures below 400°C, the stress–strain relationships may be extended by the strain hardening option, provided local instability is prevented and the ratio f_u / F_y is limited to 1.25. The ultimate stress f_u at elevated temperatures may be determined using the following:

$$f_u = 1.25 f_y \qquad \text{for } 20\ ^\circ\text{C} \leq \text{T} \leq 300\ ^\circ\text{C} \qquad (5.9\text{a})$$
$$f_u = f_y (2.0 - 0.0025\text{T}) \qquad \text{for } 300\ ^\circ\text{C} < \text{T} < 400\ ^\circ\text{C} \qquad (5.9\text{b})$$
$$f_u = f_y \qquad \text{for } 400\ ^\circ\text{C} \leq \text{T} \leq 1200\ ^\circ\text{C} \qquad (5.9\text{c})$$

Where:
    T = Steel temperature, °C

For strains ε higher than 2%, the stress–strain relationships allowing for strain hardening could be determined as follows:

$$f = f_y + 50\ (\varepsilon - 0.02)\ (f_u - f_y) \qquad \text{for } 0.02 < \varepsilon < 0.04 \qquad (5.10\text{a})$$
$$f = f_u \qquad \text{for } 0.04 \leq \varepsilon \leq 0.15 \qquad (5.10\text{b})$$
$$f = 20\ f_u\ (0.2 - \varepsilon) \qquad \text{for } 0.15 < \varepsilon \leq 0.20 \qquad (5.10\text{c})$$
$$f = 0 \qquad \text{for } \varepsilon \geq 0.20 \qquad (5.10\text{d})$$

The effect of strain hardening could be considered only if the analysis is based on advanced calculation models (see Section 5.3.2). This is allowed only if it is proven that local failures (local buckling, shear failure, spalling, etc.) do not occur because of increased strains.

The stress–strain relationships of Equations 5.8 are valid for increasing temperature (heating) histories. For decreasing temperature (cooling) histories, these relationships can be used as a sufficiently precise approximation.

Figure 5.7 illustrates the stress-strain relationships described by Equations 5.8 through 5.10.

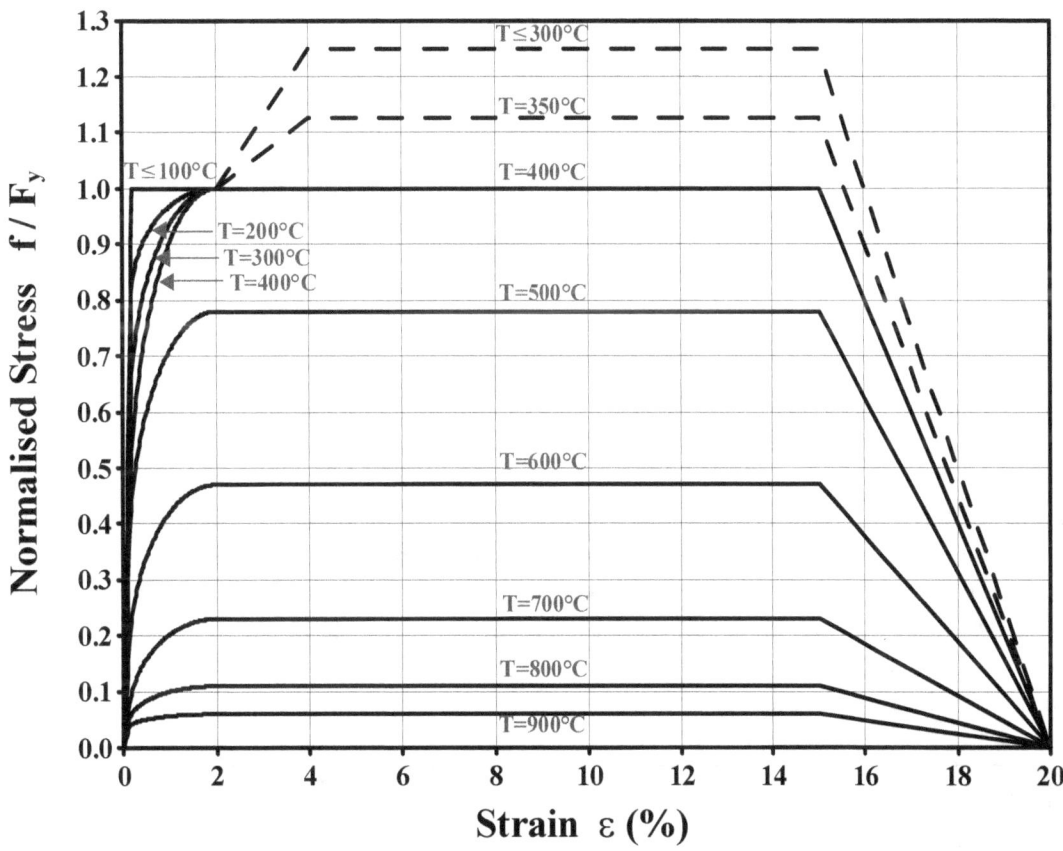

FIGURE 5.7. Stress–Strain Relationships for Structural Steel at Elevated Temperatures (ECCS 2001)

The residual yield stress f_{yR} and tensile strength f_{uR} of steel heated to a maximum temperature T_{max} and having cooled down to the ambient temperature of 20°C may be determined as follows:

$$f_{yR} = F_y \qquad \text{for } 20\ ^\circ C \leq T_{max} \leq 800\ ^\circ C \qquad (5.11a)$$
$$f_{uR} = F_u \qquad \text{for } 20\ ^\circ C \leq T_{max} \leq 800\ ^\circ C \qquad (5.11b)$$
$$f_{yR} = 0.9\ F_y \qquad \text{for } 800\ ^\circ C < T_{max} \leq 1200\ ^\circ C \qquad (5.11c)$$
$$f_{uR} = 0.9\ F_u \qquad \text{for } 800\ ^\circ C < T_{max} \leq 1200\ ^\circ C \qquad (5.11d)$$

5.2.1.2.2 Stress-Strain Relationship - NIST Approach

In support of the World Trade Center collapse investigation, NIST characterized the high-temperature stress-strain behavior of seventeen different steels recovered from the collapse site. The steels in this group represent construction steels with specified yield strengths in the range of 36 ksi to 100 ksi (248 MPa to 689 MPa). The data set included examples of plates, heavyweight shapes, and lightweight shapes. Full stress-strain data as a function of temperature were obtained for nine of the steels. In addition, a survey of the technical literature was conducted to obtain existing data on stress-strain behavior of other construction steels. This background information is reported in NIST NCSTAR 1-3D (Luecke 2005). These data were subsequently reanalyzed to

133

develop general models for the tensile stress-strain behavior of construction steels as a function of temperature.

Modulus of Elasticity

Using stress-strain data for three steels recovered from the WTC towers collapse, the following expression was obtained for modulus of elasticity (slope of linear portion of σ-ε curve up to the proportional limit) of structural steel:

$$E(T) = e_0 + e_1 T + e_2 T^2 + e_3 T^3 \qquad (5.12)$$

where,

e_0	=	206 Gpa	=	29900 ksi
e_1	=	0.043 GPa/°C	=	-6.3 ksi/°C
e_2	=	-3.5×10^{-5} GPa/°C^2	=	-5×10^{-3} ksi/°C^2
e_3	=	-6.6×10^{-8} GPa/°C^3	=	-9.6×10^{-6} ksi/°C^3

The constant term, e_0, was taken as the average value of modulus of elasticity of three steels tested at room temperature. Equation 5.12 is plotted in Figure 5.8. Also, plotted are experimental values for two 55 ksi (379 MPa) plates and one 50 ksi (344 MPa) plate tested by NIST and for results reported for nine series of tests obtained from the technical literature (see NIST NCSTAR 1-3D). Values obtained using the Eurocode 3 reduction factors (Table 5.1) are also plotted as a comparison to the NIST model.

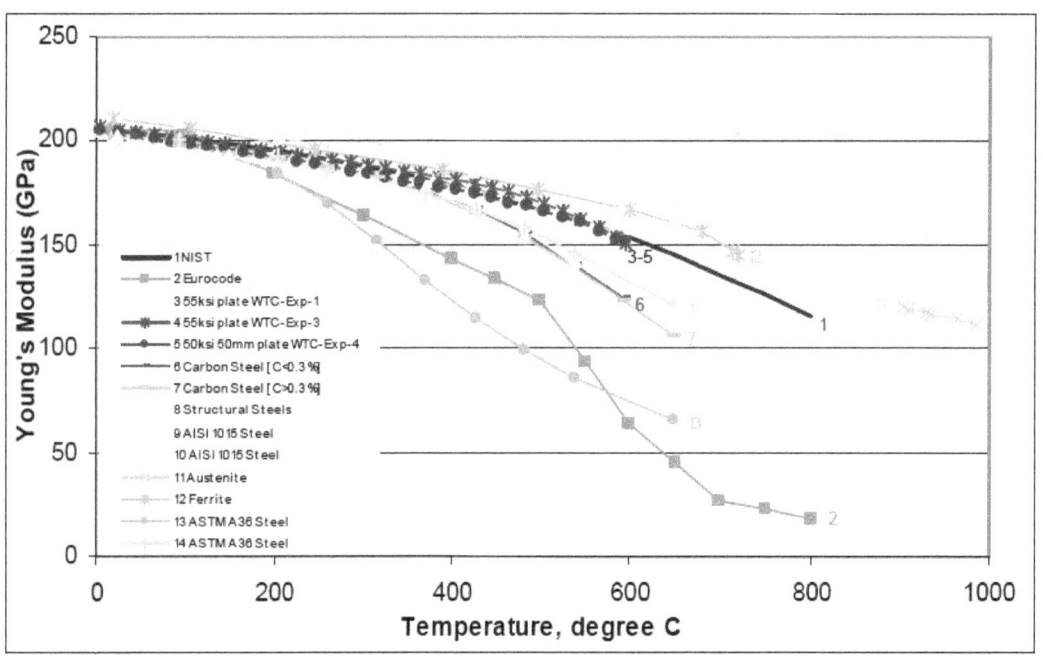

FIGURE 5.8. Comparison of Modulus of Elasticity of Structural Steels at Elevated Temperatures (See Table 5.6 for references).

Stress-Strain Relationship - True Stress-True Strain

A simple, phenomenological model for the relation between true stress σ and true strain ε is the power-law strain-hardening expression:

$$\sigma = K\varepsilon^n \tag{5.13}$$

The above equation can be expressed as a function of temperature using the following expressions for K and n:

$$K = \left(k_3 + k_4 F_y \right) \exp\left(-\left(\frac{T}{k_2} \right)^{k_1} \right) \tag{5.14a}$$

$$n = \left(n_3 + n_4 F_y \right) \exp\left(-\left(\frac{T}{n_2} \right)^{n_1} \right) \tag{5.14b}$$

where, F_y is the room temperature yield strength of steel.

The remaining parameters in Equations 5.14a and 5.14b can be determined by best fit to experimental data. Values are given in Table 5.2 for the steels tested by NIST (F_y of 36 ksi to 100 ksi) and should give good results for steels with room temperature yield strengths in the range of 36 ksi to 65 ksi.

TABLE 5.2. Parameters for Equation 5.12.

Parameter	Value	Units
k_1	4.92	
k_2	575	°C
k_3	734	MPa
k_4	0.315	
n_1	4.51	
n_2	637	°C
n_3	0.329	
n_4	-4.23×10^{-4}	MPa^{-1}

An example is provided below for an ASTM A572 Grade 50 steel with a measured room temperature yield strength of 52.5 ksi (362 MPa). In this example, the true stress is computed at three temperatures (20 °C, 400 °C, and 600 °C) using Equations 5.13 and 5.14 and coefficients given in Table 5.2. The computed values are shown in Table 5.3. Figure 5.9 shows a plot of these computed values along with experimental data obtained by NIST.

TABLE 5.3. True stress-true strain values for an A572 Grade 50 structural steel using Equations 5.13 and 5.14.

True Stress Calculation						
	20 C		400 C		600 C	
True Strain	σ_{True} (MPa)	σ_{True} (ksi)	σ_{True} (MPa)	σ_{True} (ksi)	σ_{True} (MPa)	σ_{True} (ksi)
0.001	252	36.5	245	35.5	140	20.3
0.01	377	54.7	350	50.8	169	24.6
0.02	426	61.8	390	56.6	179	26.0
0.03	458	66.4	416	60.3	185	26.9
0.04	481	69.8	435	63.0	190	27.5
0.05	501	72.6	450	65.3	193	28.0
0.1	566	82.0	501	72.7	205	29.7
0.15	607	88.1	534	77.4	212	30.7
0.2	639	92.7	558	81.0	217	31.4
0.25	665	96.4	578	83.8	221	32.0

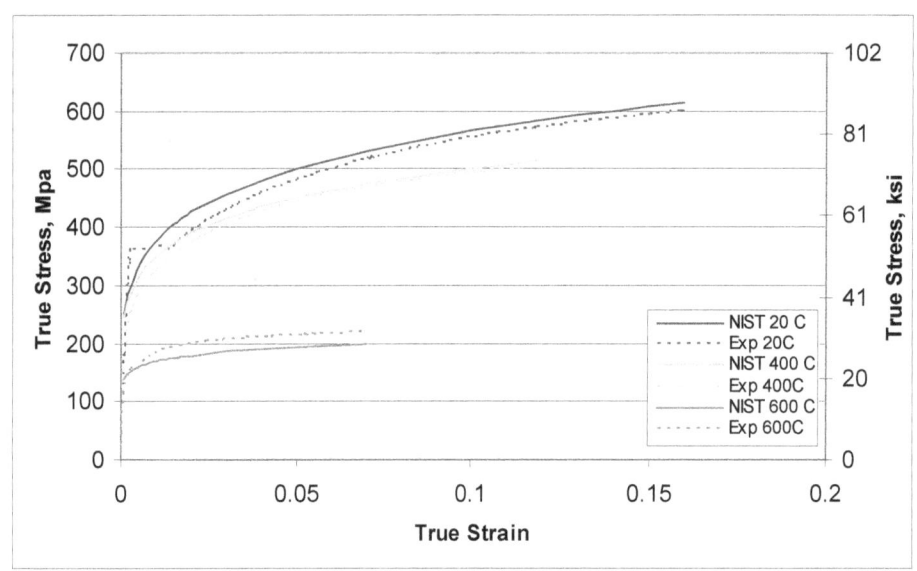

FIGURE 5.9. Comparison of true stress vs. true strain as predicted using the NIST approach and NIST experimental data for ASTM A572 Gr. 50 structural steel.

Note that the phenomenological model used here cannot reproduce a yield plateau that is observed in uniaxial tensile tests at room temperature. However, such stress-strain behavior is generally not evident in hot rolled sections with residual stresses. Further, a yield plateau is not observed at elevated temperatures.

Engineering Stress-Engineering Strain
Engineering stress and engineering strain data are often of more interest to engineers than true stress and true strain data. The following expressions can be used to convert true stress and true strain to engineering stress and engineering strain:

136

$$\sigma_E = \frac{\sigma}{\exp(\varepsilon)} \tag{5.15a}$$

$$\varepsilon_E = \frac{\sigma}{\sigma_E} - 1 \tag{5.15b}$$

As an example, consider true stress and true strain values of 88.1 ksi (607 MPa) and 0.15, respectively, corresponding to 20 °C as shown in Table 5.3. The engineering stress and engineering strain are obtained by substitution into Equations 5.15a and 5.15b, giving

$$\sigma_E = 88.1 / \exp(0.15) = 75.8 \text{ ksi}$$
$$\varepsilon_E = 88.1 / 75.8 - 1 = 0.16$$
$$\text{or}$$
$$\sigma_E = 607 / \exp(0.15) = 522 \text{ MPa}$$
$$\varepsilon_E = 607 / 522 - 1 = 0.16$$

The engineering stress vs. engineering strain values for an A572 Grade 50 structural steel as computed above are plotted in Figure 5.10 for 20 °C, 400 °C and 600 °C. Also plotted in the figure are the corresponding engineering stress and strain data obtained from experiments and from the Eurocode 3 model. Note that, for the Eurocode model, values for temperatures below 400 °C were computed assuming strain hardening using the equations provided in Annex A of Eurocode 3.

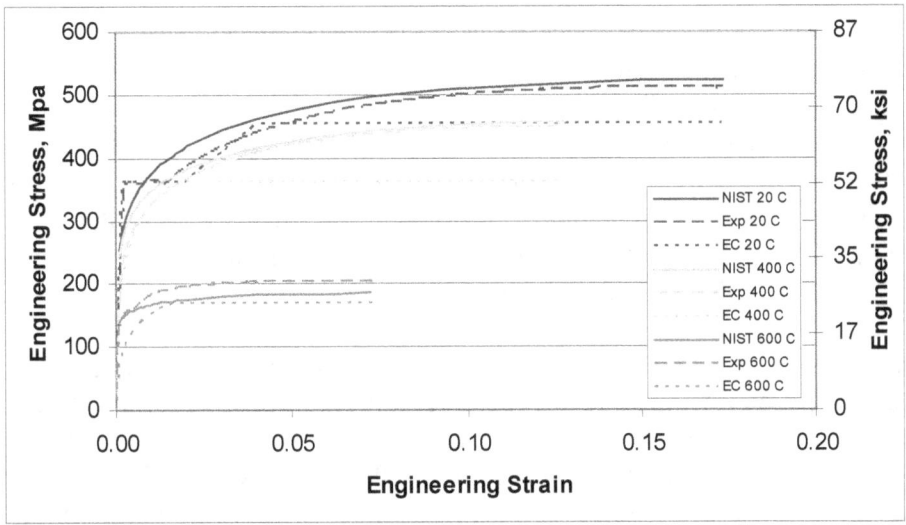

FIGURE 5.10. Comparison of engineering stress vs. engineering strain as predicted using the NIST approach, the Eurocode equations, and NIST experimental data for ASTM A572 Gr. 50 structural steel.

The stress-strain curves shown in Figures 5.9 and 5.19 terminate at the tensile strength measured in the tensile tests from which the stress-strain curves were developed. Beyond the tensile strength, the strain is not uniform (both true strain and engineering strain) since necking occurs in a tensile test specimen.

Yield Stress

The stress-strain relationship (Equation 5.13) is a power-law work-hardening model that was fitted to a large range of strain (i.e., from zero to 0.15 or greater). This model works well for large strains, but does not capture the linear elastic behavior of steels at small strains. For analysis purposes, the linear elastic portion of the stress-strain curve can be modeled separately up to the yield stress (as defined by a 0.002 strain offset), and the power-law work-hardening model can be used to provide stress-strain data beyond the yield stress.

Since the yield stress varies with temperature, a relationship is required for the yield stress at elevated temperatures. Equation 5.16 (see Luecke 2005) gives the ratio of the yield stress at an elevated temperature to the yield stress at room temperature (where elevated temperature yield stresses are also defined at a 0.002 offset strain).

$$R = \frac{F_y(T)}{F_y(20^0C)} = \left(1 - A_2\right)\exp\left(-\frac{1}{2}\left(\frac{T}{s_1}\right)^{m_1} - \frac{1}{2}\left(\frac{T}{s_2}\right)^{m_2}\right) + A_2 \tag{5.16}$$

Figure 5.11 is a plot of the normalized yield stress ratio for data from WTC structural steels and from the technical literature. Tests conducted by NIST followed ASTM E21, *Standard Test Methods for Elevated Temperature Tension Tests of Metallic Materials*, using a loading rate of $d\varepsilon/dt = 0.005/min \pm 0.002/min$. Data from the technical literature are plotted for only those steels tested at strain rates $d\varepsilon/dt \leq 0.007/min$, the upper limit for tests that conform to the requirements of ASTM E21. The solid (blue) line is given by Equation 5.16, where $A_2 = 0.075$, $m_1 = 8.07$, $m_2 = 1.0$, $s_1 = 635$ °C, and $s_2 = 539$ °C.

The Eurocode defines separate curves for the proportional limit and the yield stress. However, the yield stress at elevated temperatures is defined at a 0.02 offset strain in the Eurocode. The upper (blue) dashed line in Figure 5.11 is the Eurocode 3 yield stress at 0.02 offset strain normalized to the room-temperature yield stress at 0.002 offset strain. The lower (red) dashed line in Figure 5.11 is the Eurocode 3 proportional limit, normalized to the room temperature yield stress at 0.002 offset strain.

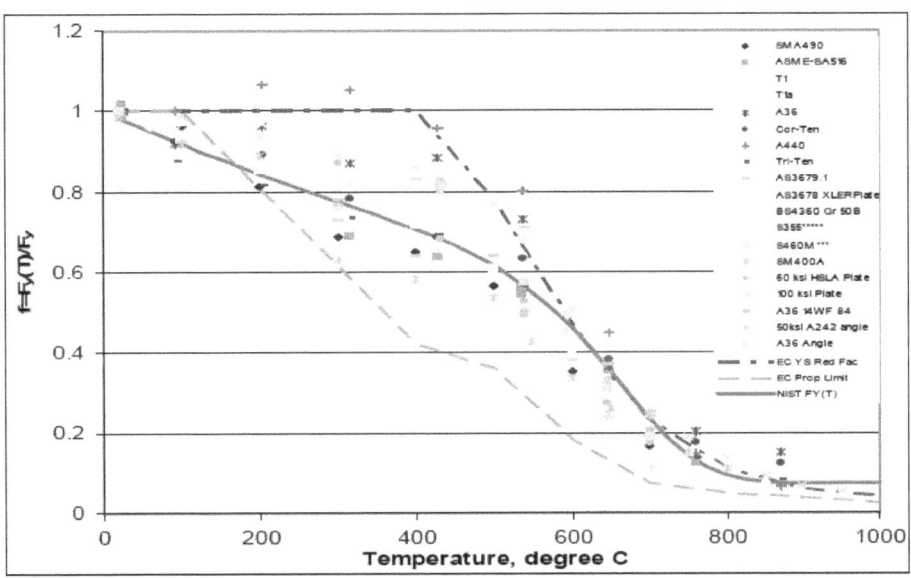

FIGURE 5.11. Elevated temperature yield stress normalized by room temperature yield stress (see Table 5.7 for references).

For analysis purposes, a temperature dependent stress-strain relationship can be developed based on the relationships for the elastic modulus (Equation 5.12), yield stress (Equation 5.16), and true stress-true strain (Equations 5.15a and 5.15b). An example of a stress-strain relationship based on these relationships for an ASTM A572, Grade 50 steel is shown in Figure 5.12. The modulus of elasticity (Equation 5.12) and yield stress (the room temperature yield stress multiplied by R from Equation 5.16) were determined for 20 °C, 400 °C, and 600 °C. The strain at the yield stress could be slightly offset from the 0.002 value, depending on where the modulus of elasticity and yield stress intersect. Discrete strains greater than that at the yield stress were selected (for a tabular input format) and the corresponding stresses were determined with Equations 5.13 and 5.14. These true stress and true strain values were converted to engineering stress and strain using Equations 5.15a and 5.15b. Figure 5.12 (a) is a plot of the elevated temperature relationship up to a strain of 0.2, and Figure 5.12 (b) shows the lower portion of the curve to a strain of 0.05.

By adjusting the chemistry of a steel, one can obtain improved properties under fire exposure. Such steels, sometimes called "fire-resistive steels" have been shown to produce a different decay curve than that in the above illustration. The reduction in yield stress with temperature for a given fire-resistive steel can be obtained from Equation 5.16 using parameters established from elevated temperature tension test data using, for example, ASTM E 21 standard test methods.

139

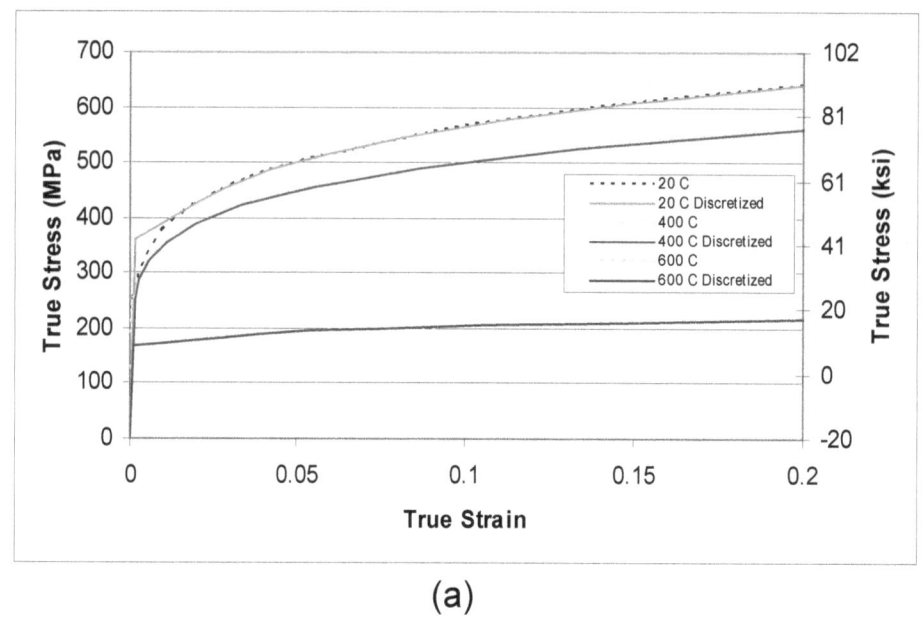

(a)

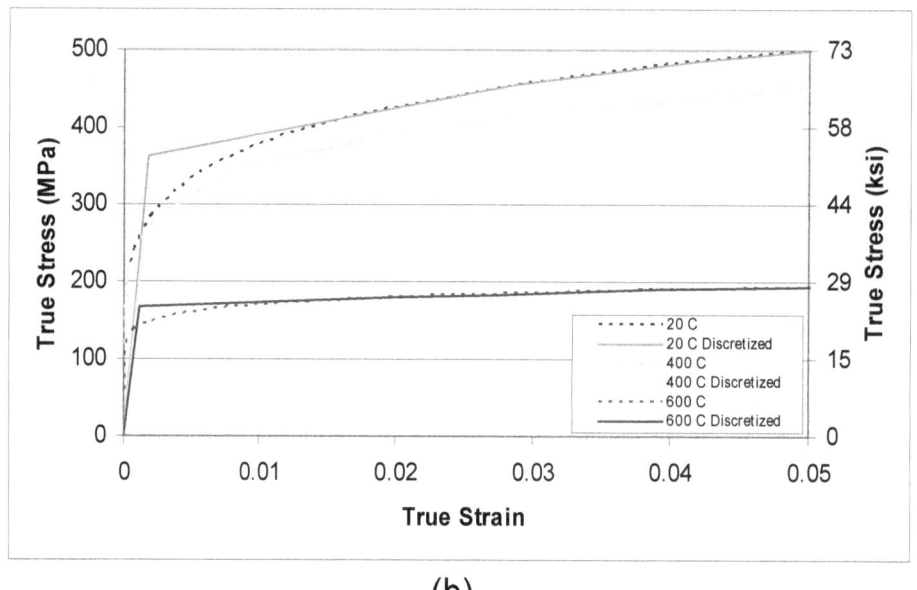

(b)

FIGURE 5.12. Computed stress-strain curves for an ASTM A 572 Grade 50 steel.

Strain rate effects can be addressed by adding a strain-rate sensitivity term, m(T), to Equation 5.13:

$$\sigma = \left(\frac{\dot{\varepsilon}}{\dot{\varepsilon}_0}\right)^{m(T)} K\varepsilon^n \qquad\qquad 5.17$$

where $\dot{\varepsilon}_0$ is the strain rate under which the original data were developed.

5.2.2 Fire Protection Materials

Fire protection materials and systems are designed to delay the temperature rise in structural steel, usually through one or more of the following mechanisms:

- Low thermal conductivity
- High effective heat capacity
- Heat-absorbing physical reactions (e.g., transpiration, evaporation, sublimation, ablation) or chemical reactions (e.g., endothermic decomposition, pyrolysis)
- Intumescence, i.e., formation of a thicker insulating foam upon heating
- Reradiation

The common fire protection materials that provide one or more of these mechanisms function generally in one of three ways (in addition to radiation shielding):

1. Mostly insulating—These are low-conductivity, lightweight, spray-applied fire-resistive materials (SFRM); mineral fiberboard products; and ceramic wool wraps.
2. Energy absorbing—These are most commonly gypsum-based or concrete-based products, each of which releases water of crystallization when exposed to high temperatures.
3. Intumescent—Applied as multi-layer paint, these coating systems expand upon exposure to high temperatures, forming an insulating layer. They are traditionally more expensive but provide many benefits, including reduced weight, durability, aesthetic appeal, good adhesion, and the option for off-site application that saves construction time on site.

Fire protection materials could be classified as organic or inorganic, depending on the chemistry of their major components. Therefore, SFRM are often referred to as inorganic systems, while intumescent coatings are often referred to as organic systems.

SFRM products today are the most common type of fire protection for structural steel. Gypsum board and mineral fiber board products, intumescent coatings, and ceramic wool wraps are common alternatives. Plaster, clay tile, concrete, and masonry enclosures, commonly used to protect steel a few decades ago, are not very common in modern fire protection. For the most comprehensive list of commercial products used in fire-resistant designs, refer to the Underwriters Laboratories (UL) Directory (*Fire* 2004). It should also be noted that steel itself is an effective fire-protective material when used in the form of wrapping sheets (e.g., UL Designs X101, X203, X526) to provide protective and reflective shielding, or in the form of meshes, banding straps, and wire ties (e.g., UL Designs Y714, XR614, X204, X205) to help fire protection materials maintain their integrity under heat exposure. Other fire protection methods for structural steel involve rain screens (sprinklers designed to protect steel members) or filling tubular structures with concrete or water.

Table 5.4 lists the thermal properties of some common fire protection materials suggested by ECCS (1995) for simplified heat transfer calculation models (see Section 5.3.2.1). Note that the values reported are for room temperature only and thermal properties of many fire protection

141

materials in common use are known to be temperature dependent. NIST NCSTAR 1-6A, Chapter 6 provides temperature dependent properties of several SFRMs (Carino et al., 2005). More detailed discussion of common fire protection materials is provided below.

TABLE 5.4. Thermal Properties of Common Fire Protection Material (ECCS 1995)

Material	unit mass ρ_p [kg / m³]	mosture content p [%]	thermal conductivity λ_p [W / (m·k)]	specific heat c_p [J/(kg·K)]
Sprays				
- mineral fibre	300	1	0.12	1200
- vermiculite cement	350	15	0.12	1200
- perlite	350	15	0.12	1200
High-density sprays				
- vermiculite (or perlite) and cement	550	15	0.12	1100
- vermiculite (or perlite) and gypsum	650	15	0.12	1100
Boards				
- vermiculite (or perlite) and cement	800	15	0.20	1200
- fibre-silicate or fibre-calcium-silicate	600	3	0.15	1200
- fibre-cement	800	5	0.15	1200
- gypsum board	800	20	0.20	1700
Compressed fibre boards				
- fibre silicate, mineral-wool, stone-wool	150	2	0.20	1200
Concrete	2300	4	1.60	1000
Light weight concrete	1600	5	0.80	840
Concrete bricks	2200	8	1.00	1200
Bricks with holes	1000	-	0.40	1200
Solid bricks	2000	-	1.20	1200

5.2.2.1 Spray-Applied Fire-Resistive Materials (SFRM)

Most SFRM utilize either mineral fiber or cementitious materials to insulate steel from the heat of a fire. Mineral fiber and vermiculite acoustical plaster on metal lath are two of the frequently used SFRM on steel columns, beams, and joists. These popular commercial products have proprietary formulations, and, therefore, it is imperative to follow closely the manufacturer's recommendations for mixing and application.

The mineral fiber mixture combines the mineral fibers, binders (usually, Portland cement based), air, and water. Mineral fiber fire protection material is spray-applied with specifically designed equipment that feeds the dry mixture of mineral fibers and various binding agents to a spray nozzle, where water is added to the mixture as it is sprayed on the surface to be protected. In the final cured form, the mineral fiber coating is usually lightweight, essentially non-combustible, chemically inert, and a poor conductor of heat.

Cementitious SFRM are composed of a binder material mixed with aggregates. Various additives and foaming agents are also often mixed in. The common binders are calcined gypsum and Portland cement. Some formulations use magnesium oxychloride, magnesium oxysulfate, calcium aluminate, calcium phosphate, or ammonium sulfate. Common aggregates are vermiculite and perlite. Some manufacturers have substituted polystyrene beads for the vermiculite aggregate. The frequently used vermiculite acoustical plaster is a cementitious product composed of gypsum binder and perlite or vermiculite lightweight aggregates. Cementitious SFRM can be classified by their density as low-density (about 240 kg/m^3), medium-density (320 kg/m^3 to 430 kg/m^3), and high-density (640 kg/m^3 to 1280 kg/m^3) products. Higher density SFRM products (1600 kg/m^3 to 2400 kg/m^3) are essentially concrete mixtures.

Lie (1992) suggests the approximate value of k = 0.1 W/m°C for mineral fiber SFRM products with densities ρ = 250 kg/m^3 to 350 kg/m^3, and cementitious SFRM products with densities ρ = 800 kg/m^3 to 1000 kg/m^3. Ruddy et al. (2003) suggest the following representative thermal properties for lightweight SFRM (both mineral fiber and cementitious):

c = 0.754 kJ/kg°C
k = 0.135 W/m°C
ρ = 293 kg/m^3

The durability characteristics and on-site application quality of SFRM are regulated by the following standards:

- ASTM E605, "Standard Test Methods for Thickness and Density of Sprayed Fire-Resistive Material Applied to Structural Members"
- ASTM E736, "Standard Test Method for Cohesion/Adhesion of Sprayed Fire-Resistive Materials Applied to Structural Members"
- ASTM E759, "Standard Test Method for Effect of Deflection on Sprayed Fire-Resistive Material Applied to Structural Members"
- ASTM E760, "Standard Test Method for Effect of Impact on Bonding of Sprayed Fire-Resistive Material Applied to Structural Members"
- ASTM E761, "Standard Test Method for Compressive Strength of Sprayed Fire-Resistive Material Applied to Structural Members"

5.2.2.2 Intumescent Coatings

An intumescent coating has the appearance of a thick film or paint. When exposed to a fire, it chars, foams, and expands significantly in thickness. To retain this insulating layer, reinforcing is sometimes required at sharp corners, such as the flange tips of a wide-flange shape.

The intumescent mechanism involves the interaction of four types of compounds (Yandzio, Dowling, and Newman 1996):

1. Inorganic acid, or material yielding an acid at temperatures of 100 to 300 °C. Ammonium polyphosphates, yielding phosphoric acid, are common for this purpose.
2. A polyhydric compound rich in carbon (such as starch) that reacts with the acid to form carbonaceous char.
3. A spumific agent that decomposes to liberate large volumes of non-combustible gases (including carbon dioxide, ammonia, and water vapor). Production of theses gases causes the carbonaceous char to foam and expand into a thick (up to 100 times thicker that original) protective layer.
4. A binder or resin that softens at a predetermined temperature and helps preventing gases from escaping.

A review of chemical and physical phenomena that occur within intumescent coatings upon heating, and various approaches to model intumescent behavior, has been provided by Butler (1997).

Two distinct categories of intumescent coating are manufactured: water or solvent based (also referred to as intumescent paints or thin film coatings), and epoxy based (also referred to as mastics or thick film coatings). Water and solvent-based coatings are thinner (usually up to 5 mm) and mostly intended for controlled environments inside buildings, although some systems are available for external exposures. Epoxy-based coatings are thicker (up to 45 mm) and are mostly used for petro-chemical installations.

In many instances, the intumescent coating is actually a system of multiple coats with different properties and functions. The base coat (or primer coat) is formulated to provide a strong bond to the steel substrate, while the top coat (or the sealing/decorative coat) is formulated to provide a durable aesthetically appealing finished surface. The intermediate layer of the fire protective intumescent material is usually applied in multiple coats to achieve the desired protection thickness, allowing sufficient time for each coat to dry before applying the next coat. Therefore, depending on the design thickness, the application of intumescent coatings could be a lengthy and costly process. Cost-effective solutions are often achieved through the over-design of structural steel, i.e., increasing the size (the W/D ratio) of steel sections, that results in the reduction of the design thickness of intumescent protection and the reduction in the overall cost of construction.

All intumescent coating products are proprietary and their thermal properties have rarely been reported in the open literature, although special proprietary software (Lawson, Oshatogbe, and Newman 2002) provides predictions of steel section (protected by intumescent coating) temperatures under standard fire exposure. Hamins (1998) reported the dry density of eight different intumescent coatings ranging within 1.1 g/cm^3 to 1.6 g/cm^3. After 30 minutes of heating in this study, the measured expansion factors for the char ranged between 2 and 56, and the measured mass loss (attributed to vaporization) ranged between 38% and 66%, suggesting a reduction in char density (compared to the original coating density) of up to about 100 times. An earlier study by Anderson, Ketchum, and Mountain (1988) reported measurements of the thermal conductivity of 13 different intumescent chars that averaged 0.096 ($\pm$0.033) W/m°C.

The only public document in the United States to regulate the on-site application and quality of intumescent coatings is the AWCI Technical Manual 12-B (1998). Off-site (workshop) application of intumescent coatings, with subsequent on-site erection of coated structural steel, is relatively rare in the United States. Off-site application has been more popular in the United Kingdom (Yandzio, Dowling, and Newman 1996).

5.2.2.3 Gypsum Board Products

Gypsum board consists of a non-combustible core (usually, of not less than 65 % gypsum) and paper laminated surfaces to form flat sheets available in a range of sizes. Special fire-resistant types of gypsum board are usually used for fire protection. Type X and "Type C" (because of the letter "C" usually appearing in the designation of the associated products) board are the fire-resistant products most widely used in North America. These products have a specially formulated gypsum core in order to provide a greater fire resistance (of building systems) than regular gypsum board of the same thickness.

Type X gypsum board is defined (ASTM C36-97) as the board that would provide the following minimum fire resistance ratings for the assemblies described:

- One hour for a non-load-bearing cold formed steel frame wall with galvanized 92x35x0.5 mm steel studs spaced 600 mm o.c. and a single layer of 15.9 mm thick gypsum board on each side. The 1200 mm wide gypsum board sheets should be attached to the studs using 25 mm long drywall screws spaced 200 mm o.c. along edges and ends, and 300 mm o.c. along intermediate studs. All joints should be oriented parallel to, and located over, studs and staggered on opposite sides of the assembly.
- Two hours for a non-load-bearing cold formed steel framed wall with galvanized 64x35x0.5 mm steel studs spaced 600 mm o.c. and two layers of 12.7 mm thick gypsum board on each side. The gypsum board sheets should be 1200 mm wide. The base layer sheets should be attached to the studs using 25 mm long drywall screws spaced 300 mm o.c. along board edges, ends, and along intermediate studs. Base layer joints should be oriented parallel to, and located over, studs and staggered on opposite sides of the assembly. The face layer sheets should be attached using 41 mm long drywall screws spaced 300 mm o.c. along board edges, ends, and along intermediate studs. Face layer joints should be oriented parallel to, and located over, studs, offset 600 mm from base layer joints, and staggered on opposite sides of the assembly.

Type C gypsum board does not have an associated product standard. Type C products are advertised by gypsum board manufacturers as exhibiting superior fire performance characteristics (compared to Type X products) and exceeding the ASTM requirements for Type X gypsum board.

Gypsum (calcium sulphate) dihydrate, $CaSO_4 \cdot 2H_2O$, the major component of the gypsum board core, contains chemically bound water of crystallization (about 21% by weight) in addition to a small amount of free moisture. The fire retarding property of gypsum board products derives primarily from this water content. When the gypsum board is exposed to fire, the water of crystallization is gradually released and evaporated, consuming large amounts of energy in the process and delaying heat transmission through the board. Therefore, gypsum board effectively acts as a fire barrier until most of its water content is driven out. As shown in Figure 5.13, the temperature of the protected steel directly behind the fire-exposed gypsum board remains around the boiling temperature of water (100 °C) for the duration of the delay. The dehydration (release of water of crystallization) of gypsum dihydrate proceeds in two phases. The first phase, known as calcination, is associated with the formation of gypsum hemihydrate (semihydrate), $CaSO_4 \cdot 1/2H_2O$, at temperatures in the range of 80 °C to 150 °C. The second phase of dehydration occurs at higher temperatures when the gypsum hemihydrate is transformed to gypsum anhydrite, $CaSO_4$. In the case of fire near gypsum products, calcination of the core starts at the fire-exposed surface and penetrates through the thickness of the board. The progress of calcination is retarded by the layer of calcined gypsum on the fire-exposed surface, which adheres well to the inner uncalcined layers.

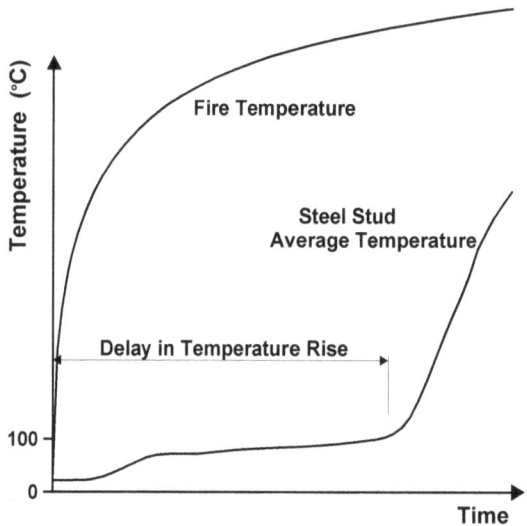

FIGURE 5.13. Typical Temperature History of a Steel Member Protected with Gypsum Board

The thermo-physical properties of the gypsum board (Alexander 1982) could vary depending on the composition of the core and depending on the methods used to derive the properties. Figures 5.14 and 5.15 illustrate the typical variation of the specific heat and the thermal conductivity, respectively, of the gypsum board core with temperature. The plots reflect the expressions proposed by Sultan (1996) based on tests conducted on Type X gypsum board specimens. Specific heat measurements were carried out at a heating rate of 2 °C/min. The dehydration of gypsum resulted in the two peaks that appear in the specific heat curve at temperatures around 100 °C and 650 °C.

The Gypsum Association (Gypsum 1998) lists typical mechanical properties, at room temperature, for some North American gypsum board products. Little is known about the mechanical properties of the gypsum board at elevated temperatures because these properties are

146

difficult to obtain experimentally. However, the tensile strength of the gypsum board has a significant effect on its performance in fire. Gypsum board shrinks significantly at elevated temperatures (up to 1.5% by the time it reaches 600 °C). When this shrinkage is restrained, considerable tensile stresses build up in the board, causing cracks within the field of the board and/or near the attachment screws.

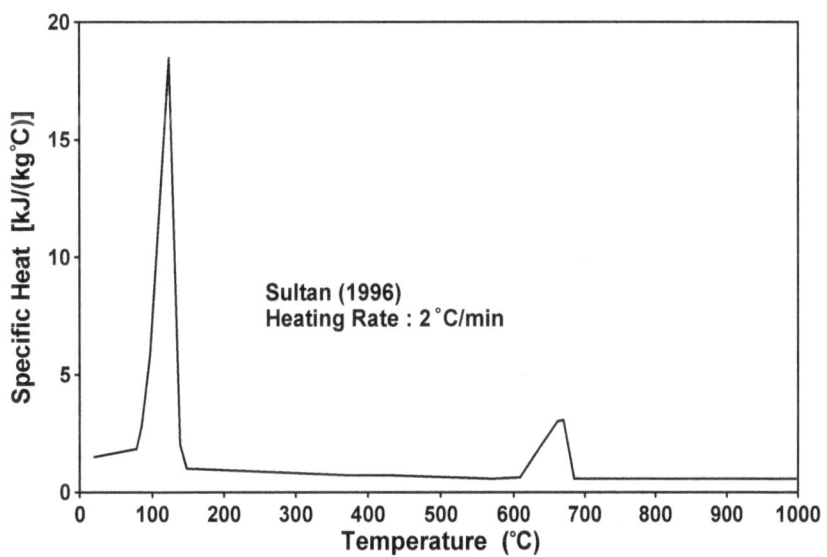

FIGURE 5.14. Specific Heat of Type X Gypsum Board Core

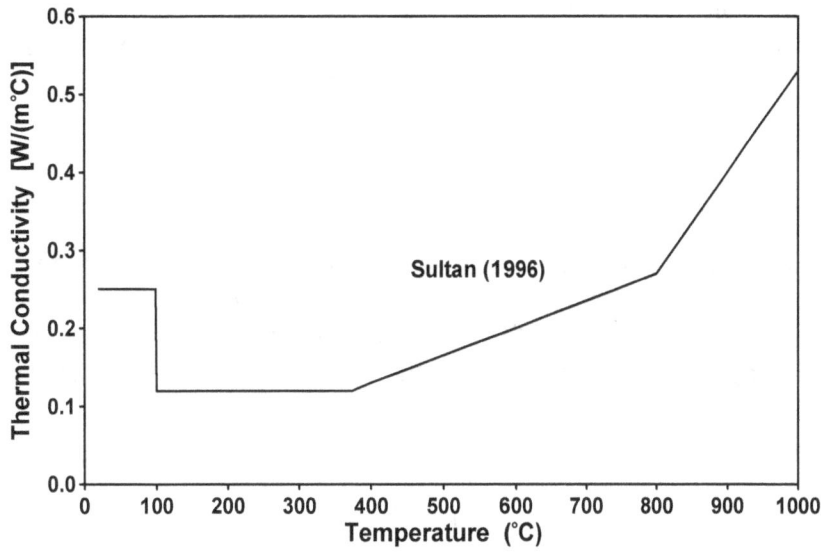

FIGURE 5.15. Thermal Conductivity of Type X Gypsum Board Core

Even dehydrated gypsum board acts as a shield against fire flames and associated radiation emissions. The enhanced fire performance of Type X and Type C gypsum board derives primarily from the improved ability of these products to stay in place for an extended period of time. This improved ability is usually achieved through reduced shrinkage and increased strength characteristics of the gypsum core. Special additives are used in the proprietary formulations of fire-resistant boards to reduce their shrinkage (often clay and vermiculite) at elevated temperatures. Reinforcement fibres (usually glass fibres) are added to improve the mechanical properties of the boards. Type X and Type C products are usually of higher density than regular gypsum board.

5.2.2.4 Fibrous Board and Mat Products

Although mineral fiberboard and similar mat products are usually more expensive than SFRM, they are relatively easy to install and maintain since no wet processes are involved. They are often used in retrofit applications and projects where speed and dry process are of importance. Mineral fiberboard fire protection products are usually semi-rigid, while the mat products are fully flexible. These proprietary products are often supplied with an outer sheathing (aluminum foil or similar). They are fixed to structural steel using a wide variety of methods: steel weld pins with non-return washers, wire ties, special nails, screws, and, sometimes, bonding agents. Some protection systems also involve sealing of the joints with special tape.

Fibrous insulation products used for fire protection of structural steel range in density from 100 kg/m^3 to 320 kg/m^3 (generally, insulations used for high-temperature applications are much denser than standard insulations used for climate protection of premises). Their heat transfer properties over the full range of fire temperatures have been rarely reported in literature. Pelanne (1978) discusses the mechanisms of heat transfer through fibrous insulations and graphically illustrates the thermal conductivity for several insulation products at mean temperatures of up to 760 °C. He points out that the thermal conductivity of lighter insulations tends to increase sharply with temperature, but this effect is much less pronounced for denser insulation products. Lie (1992) suggests an approximate temperature-independent value of $k = 0.25 \text{ W/m°C}$ for mineral wool boards with densities of $\rho = 120 \text{ kg/m}^3$ to 150 kg/m^3.

5.2.2.5 Concrete and Masonry

For the purposes of determining the fire resistance ratings of structural steel protected with concrete or masonry, Tables 720.5.1(2) and 720.5.1(3) of the IBC (2000) and Tables 5-1 through 5-3 of ASCE 29-99 suggest the following temperature-independent thermal properties (c_c is specific heat, m_c is moisture content, k_c is thermal conductivity and ρ_c is density):

For normal weight concrete:

$c_c = 0.84 \text{ kJ/kg°C}$ $\quad$ $m_c = 4 \%$ $\quad$ $k_c = 1.64 \text{ W/m°C}$ $\quad$ $\rho_c = 2323 \text{ kg/m}^3$

For lightweight concrete:

$c_c = 0.84 \text{ kJ/kg°C}$ $\quad$ $m_c = 5 \%$ $\quad$ $k_c = 0.61 \text{ W/m°C}$ $\quad$ $\rho_c = 1762 \text{ kg/m}^3$

For concrete masonry:

$c_c = 0.84$ kJ/kg°C, $m_c = 0$ %, while k_c values range from 0.36 W/m°C to 1.45 W/m°C, depending on the density (ρ_c ranging from 1281 kg/m^3 to 2403 kg/m^3)

For clay masonry:

$c_c = 1.00$ kJ/kg°C, $m_c = 0$ %, and k_c values range from 2.16 W/m°C to 3.89 W/m°C, depending on the density (ρ_c ranging from 1922 kg/m^3 to 2082 kg/m^3)

5.3 DESIGN PROCEDURES

The design of structural steel for fire resistance usually falls in one of the two broad categories of relevant design procedures: design by qualification testing (sometimes referred to as prescriptive methods), discussed in Section 5.3.1, and the more sophisticated design by engineering analysis (sometimes referred to as equivalency methods or performance-based methods) (see Section 5.3.2). While the overwhelming majority of U.S. fire-resistant design decisions are based on qualification testing, performance-based designs are increasingly gaining popularity and acceptance in the last few decades, especially in complex and high-profile projects, such as airports (Baldassarra and Romine 1988), museums and universities (Lane 2000), arts centers (Chen and Gemeny 2004), and other public buildings (Siu 2005).

5.3.1 Design by Qualification Testing

Qualification fire resistance testing in accordance with ASTM E119 (or very similar standards, UL 263 and NFPA 251) is used extensively in the United States to satisfy building code requirements for fire resistance. As it relates to structural steel, AISC Steel Design Guide 19 (Ruddy et al. 2003) and *Facts for Steel Buildings: Fire* (Gewain, Iwankiw, and Alfawakhiri 2003) provides a detailed discussion of building code requirements and approved fire-resistant design methods for steel-framed floors and roofs, and structural steel beams, columns and trusses, and connections.

Fire-resistant construction assemblies (walls, floors, roofs) and elements (beams, columns) that perform satisfactorily in standard fire resistance tests (ASTM E119, UL 263, NFPA 251) are documented (along with their respective hourly ratings) in building codes, standards, test reports and special directories of testing laboratories. Over the years, a considerable amount of accumulated test data has allowed the standardization of many fire-resistant designs involving generic (non-proprietary) materials such as steel, wood, concrete, masonry, clay tile, Type X gypsum wallboard, and various plasters. These generalized designs and methods are documented in building codes and standards, e.g., in IBC Sections 719 and 720 or in the ASCE/SFPE 29 standard, with detailed explanatory figures, tables, formulas, and charts. Fire-resistant designs that incorporate proprietary (pertaining to specific manufacturers and/or patented) materials are documented by test laboratories in relevant test reports and special directories. The largest single source of such proprietary fire resistance designs is the Underwriters Laboratories (UL) Directory (*Fire* 2004), updated annually. It contains a variety of designs for columns (X and Y series), walls (U and V series), floors (D and G series), roofs (P series), and beams (N and

S series). Several other accredited laboratories, such as Intertek Testing Services (ITS) and Omega Point Laboratories (OPL), also conduct standard fire-resistance tests and publish details of fire resistant designs in their directories (ITS 2004; OPL 2004).

To comply with fire resistance rating requirements specified by the building code, the designer usually selects suitable fire-resistant designs from the above sources. It should be noted, however, that listed designs must be followed in every detail to ensure the fire resistance rating. Departures from listed designs are often difficult to justify to the satisfaction of building officials. Additional testing could be required to establish fire-resistance ratings for new design configurations that have not been tested or listed before.

W/D Ratios
It has been long recognized that the rate of temperature rise in a structural steel member depends on its weight and the surface area exposed to heat. Therefore, the factor commonly used in fire resistant design is W/D, where W is defined as the weight per unit length of the steel member, and D is the inside perimeter of the fire protection, as illustrated in Figure 5.16 for columns (exposed to fire on four sides) and Figure 5.17 for beams (exposed to fire on three sides). Similar A/P factors are used in the fire-resistant design of tubular column sections, where A is the section area and P is the section perimeter. Accurate values of W/D and A/P ratios for various sections and configurations are listed in Tables 1-36 through 1-53 of the AISC LRFD Manual (*Manual* 2001) and Appendix A of the AISC Steel Design Guide 19 (Ruddy et al. 2003).

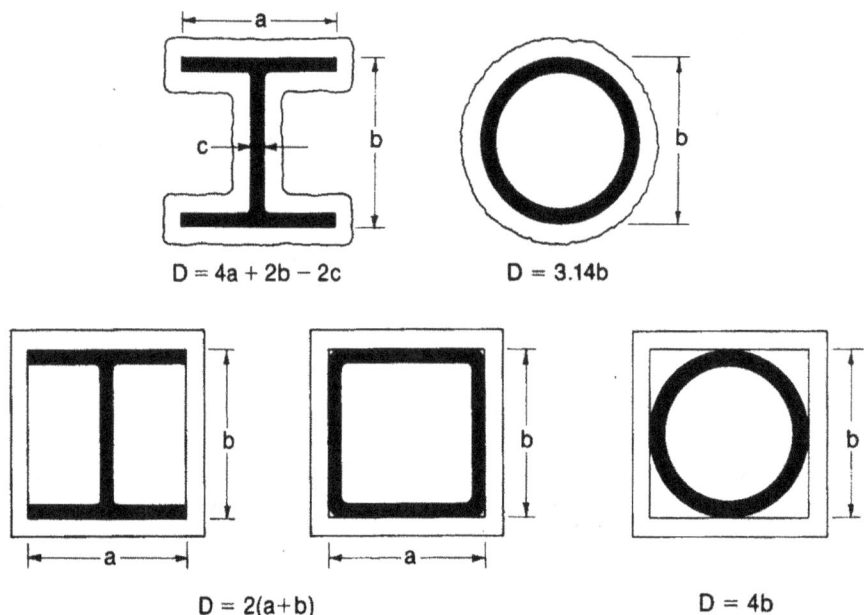

FIGURE 5.16. Perimeter D for Steel Column Sections

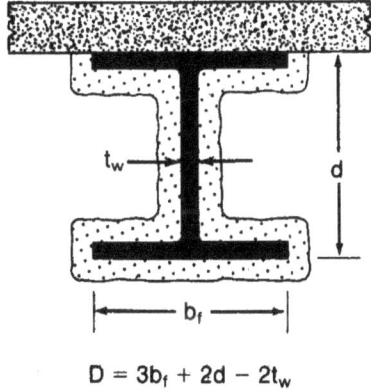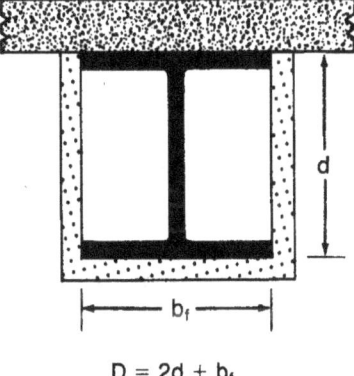

$$D = 3b_f + 2d - 2t_w \qquad\qquad D = 2d + b_f$$

FIGURE 5.17. Perimeter D for Steel Beam Sections

It should be noted that the W/D ratio approach does not apply to open web steel joists (*Design* 2003) or light-gauge cold-formed steel sections (NAHB 2004).

Structural steel sections with larger W/D ratios (or A/P ratios) experience slower rates of temperature rise in the steel and, therefore, exhibit extended fire resistance performance in fire tests and in real fire incidents. W/D and A/P ratios are often used in correlations and formulas, developed over the years from a large number of tests, for the determination of fire resistance and for the adjustment of fire protection thickness, depending on the size of the steel section.

Similar concepts are adopted in Europe and other countries; however, their correlations usually incorporate the inverted (compared to W/D) ratio H_p/A, where H_p is the inside perimeter of fire protection, and A is the steel section area.

5.3.1.1 Individual Protection

The International Building Code (2000) requires that fire-rated columns, beams, girders, trusses, and other structural members "shall be individually protected on all sides for the full length" where the structural element supports:

- More than two floors (or more than one floor and one roof), or
- A load-bearing wall, or
- A non-load-bearing wall that is more than two stories high

The requirement applies to most columns in multi-story buildings and effectively prohibits protecting more than one column in a single fire protection enclosure. This individual protection requirement also prohibits the protection of important and "critical" beams, girders, and trusses by fire-resistive ceiling membranes. However, ceiling protection (shielding more than one beam, girder, or truss by a single fire-resistant membrane) can be used for regular beam, girder, or truss

151

systems supporting one floor, or transfer beams, girders, and trusses supporting not more than two floors.

Where required for trusses, the "individual" protection is accomplished through the enclosure of the entire individual truss for its full height and length (usually by gypsum wallboard) or through the enclosure of each truss element by a spray-applied fire resistive material, intumescent coating, gypsum board, or other acceptable protection.

5.3.1.2 Structural Steel Columns

The International Building Code prescribes many column fire-resistant designs using generic (non-proprietary) materials such as concrete, masonry, plaster, and gypsum wallboard. Also, the International Building Code contains several equations (with W/D variable) and relevant tables for the calculation of fire resistance of steel columns protected with generic materials. Further, the International Building Code allows the adjustment of thickness of proprietary SFRM materials based on the W/D ratio of the column section. The ASCE/SFPE 29 standard contains very similar provisions for steel columns. In addition, the latter provides an equation for the determination of fire resistance of concrete-filled tubular steel columns (for up to 2 hours).

Fire-resistant steel column designs using proprietary materials, such as SFRM and intumescent coatings, are listed under the X and Y series designation of the UL directory. A notable exception is UL Design X107 that does not use proprietary materials; the design is for a generic 1-hour rated W-section steel column with concrete-encased web (and exposed flanges).

5.3.1.3 Steel Framed Floors and Roofs

Test standards (ASTM E119, UL 263, NFPA 251) specify dual, restrained and unrestrained, classification of fire resistance ratings for floors, roofs, beams, and girders. Floor/roof assemblies and individual beams/girders are usually tested, under maximum design load, in the restrained condition (i.e., where the thermal expansion of the test specimen is resisted by test frame), which is representative of the field conditions in most cases. The resulting restrained ratings determined during the tests are representative of the actual fire resistance of the specimens. However, in the same tests, unrestrained ratings are also determined based on certain steel temperature limits. These unrestrained ratings are sometimes used for field conditions where the thermal expansion of a certain beam or floor segment is not expected to be resisted by surrounding construction, e.g., as in the case of a cantilevered beam or floor. Table C1.1 of UL 263 (and similar tables in ASTM E119 and NFPA 251) provide guidance on the cases where unrestrained ratings are recommended. An extensive discussion of this subject is provided by Gewain and Troup (2001).

5.3.1.4 Steel Trusses

The inherently large size of truss assemblies does not allow their adequate fire resistance testing in standard furnaces. However, several conservative approaches have been developed over the years for truss fire protection. One common approach is to protect each truss element to the same level as a column of a similar or smaller section size. Another approach, sometimes used for lighter trusses, is to apply proven fire-resistant joist designs to heavier trusses. Both approaches are based on the rationale that larger/heavier truss elements would heat up more slowly than smaller column sections or lighter joists under similar fire exposures.

The column designs in the International Building Code apply equally to truss elements protected with generic (non-proprietary) materials. Further, the International Building Code specifies the method of adjustment of thickness of proprietary SFRM based on the W/D ratio of the protected steel section and the number of fire-exposed sides. For truss elements exposed to fire on four sides (vertical, diagonal, and sometimes bottom chord elements), the four-side W/D ratio should be used in this procedure. Where truss elements directly support floor or roof construction (top chord and, sometimes, bottom chord elements, as in staggered truss systems), the W/D ratio for three-side exposure is used, as for beams and girders. Similar provisions are specified in the ASCE/SFPE 29 standard.

Proprietary UL designs for columns (X and Y series) and, sometimes, walls (U and V series), floors (D and G series), roofs (P series), and beams (N and S series) can also be used for trusses. Notably, UL Design U436 is especially useful for the fire-resistant design of staggered trusses protected within a gypsum wallboard envelope.

5.3.1.5 Connections

There are no standard fire resistance tests or associated ratings for structural connections in any material (although beam-to-girder connections are sometimes included within floor assembly tests). The approach adopted by building codes for many decades has been to protect connections to the same level as the adjacent structural member with the highest fire resistance rating. For example, the International Building Code specifies that "where columns require a fire-resistant rating, the entire column, including its connections to beams and girders, shall be protected." This approach may often be satisfactory since connection failures have rarely been reported during fire incidents. In general, three factors are believed to delay temperature rise within connections during fires:

- SFRM application methods may result in higher than specified protection thickness in the corners of beam-to-column and/or beam-to-girder connections.
- The connections contain more steel material (i.e., more thermal capacity to absorb heat) as a result of the overlapping plates and/or section elements.
- The corner configuration of connections limits their exposure to fire heat radiation, while the proximity of connections to several steel members facilitates heat dissipation from the heated connection to adjacent cooler members.

It should be noted that the forces resisted by connections in real fire incidents may be significantly different from design forces. Also, multiple compartment fires and fires extending simultaneously over multiple floors may result in temperatures within connections that equal those of the connected members. Finally, the effects of thermal expansion of the structural system can lead to connection failure.

5.3.2 Design by Engineering Analysis

For a given heat exposure history (fire scenario), the engineering analysis of a steel structure would involve two major stages. First, heat transfer analysis is conducted (using simple or advanced calculation models, described in Section 5.3.2.1) to establish the temperature field history in the structure. In the second stage, the structural analysis of the heated steel structure is performed using one of the following methodologies:

- The critical temperature approach (described in Section 5.3.2.2). This is the simplest analysis methodology and involves determining critical temperatures (limiting temperatures) for various steel elements, and ensuring (adjusting section size and/or protection thickness) that these critical temperatures are not exceeded for the required time in the design fire scenario.

- Simple calculation methods (described in Section 5.3.2.3). These are generally "hand" calculation methods, although they are not necessarily simple to use (in fact, several computer programs have been developed to make these methods easier to apply). These calculation methods are based on well-established principles, such as plastic analysis of sections, and they usually are used to analyze a single member at a time. They often involve simplifying assumptions, such as neglect of thermal expansion, temperature independence, and idealization of structural boundary conditions (degrees of freedom of member ends), approximation of second-order effects, and simplified material property models.

- Advanced calculation methods (described in Section 5.3.2.4). These are generally finite element models incorporating geometrical and material nonlinearities, and they are usually used to analyze assemblies of structural components and/or entire building systems. The many assumptions and approximations in advanced calculation models are usually of a higher order of refinement than in simple calculation methods; therefore, a higher degree of accuracy is expected. Because significant expense is involved in advanced modeling and calculation, these methods are rarely used for routine design projects. Presently, they are primarily used to investigate unusual or novel structural configurations, understand frame behavior in fires, conduct parametric research studies, and to develop design aids.

5.3.2.1 Heat Transfer Analysis

Simplified heat transfer models generally assume the uniformity of temperature across the steel section at any time during the fire exposure (heating history). This assumption holds very well (due to the high conductivity of steel) for unprotected steel members and also where the structural steel member is uniformly protected on all sides and uniformly exposed to fire on all sides. For instance, steel section temperatures close to uniform are usually observed in standard fire resistance tests on unprotected or uniformly protected structural steel columns. Non-uniform temperatures could develop in steel members that have non-uniform protection and/or in cases of non-uniform fire exposures. For example, some temperature differences between the top and bottom flanges are usually observed in standard fire resistance tests on protected structural steel beams supporting concrete slabs (although this temperature difference is neglected where simplified heat transfer models are used).

154

Other significant approximations often used with simplified heat transfer models are temperature-independent (approximated) heat transfer properties for structural steel and fire protection materials (e.g., properties listed in Table 5.4) although, ideally, temperature-dependent properties could also be used with the simplified models.

Using the simplifying assumption of uniform section temperature for an unprotected structural steel member exposed to a compartment fire, the incremental temperature rise in the steel member in a short time period can be determined (Malhotra 1982) from:

$$\Delta T_s = \frac{a}{c_s \left(\dfrac{W}{D} \right)} (T_F - T_s) \Delta t \qquad (5.18)$$

where:

ΔT_s = Temperature rise in steel (°C),
D = Inner perimeter of fire protection, as defined in Figures 5.10 and 5.11 (m)
c_s = Specific heat of steel (J/(kg°C))
W = Steel section weight per unit length (kg/m)
T_F = Fire temperature (K)
T_s = Steel temperature (K)
Δt = Time step (s)

and the heat transfer coefficient, a, is determined from:

$$a = a_c + a_r \qquad (5.19)$$

where:

a_c = Convective heat transfer coefficient (W/(m²°C))
a_r = Radiative heat transfer coefficient (W/(m²°C)), defined as:

$$a_r = \frac{5.67x10^{-8}\varepsilon_F}{T_F - T_S}\left(T_F^4 - T_s^4\right) \qquad (5.20)$$

where:

ε_F = Emissivity (dimensionless)

For a standard test furnace exposure, the convective heat transfer coefficient, a_c, can be conservatively approximated as a_c = 25 W/m²°C (ECCS 2001). The conservative value for the emissivity parameter ε_F is 0.8 (ECCS 2001).

For the accuracy of temperature history calculations, the time step, Δt, should be reasonably small (e.g., not exceeding 10 seconds). It should also be noted that Equation 5.18 could be used for unprotected external steel members (members located outside the building envelope) and/or unprotected steel members exposed to localized fires. For these cases, the convective and radiative heat transfer coefficients are appropriately reduced, depending on the configuration of the considered fire scenario (ECCS 2001).

For protected structural steel members, the additional conservative simplifying assumption of outer protection surface having the temperature of the fire is employed (Malhotra 1982). Hence, for steel columns and beams with contour or box protection configurations, Equation 5.18 becomes:

$$\Delta T_s = \frac{k_p}{d_p} \left[\frac{T_F - T_s}{c_s \frac{W}{D} + \frac{c_p \rho_p d_p}{2}} \right] \Delta t \qquad (5.21)$$

where:

k_p = Thermal conductivity of the protection material (W/(m°C))
c_p = Specific heat of the protection material (J/(kg°C))
ρ_p = Density of the protection material (W/m°C)
d_p = Protection thickness (m)

and the other parameters are as defined earlier for Equation 5.18.

Equation 5.21 could be further simplified (Malhotra 1992) by conservatively neglecting the thermal capacity of the protection material:

$$\Delta T_s = \frac{k_p}{c_s d_p \frac{W}{D}} \left(T_F - T_s \right) \Delta t \qquad (5.22)$$

This approximation is appropriate where the thermal capacity of the protection material is much less than that of the steel, such that the following inequality holds:

$$c_s W/D > 2d_p \rho_p c_p \qquad (5.23)$$

Where large temperature differences across a steel section could be expected (e.g., in columns partially embedded in masonry walls, or beams partially embedded in a concrete slab), and also for complex protection and/or heat exposure configurations and any other cases where a higher degree of refinement and accuracy is desirable, the temperature field history in structural steel members can be determined using the more advanced calculation models that usually incorporate the following actions:

- Utilize finite element methods of analysis, allowing for the determination of the temperature variations within the construction assembly and the steel section.
- Where necessary, appropriately model the temperature-dependent heat transfer properties of materials.
- Where necessary, take into account the effects of cracking, spalling, or other types of high-temperature material erosion or degradation.

While many of the general-purpose finite-element programs with suitable heat transfer routines could be used to implement advanced calculations, several advanced special-purpose fire-related programs have been developed over the years for heat transfer analysis, including FIRES-T3 (SFPE 1995), TASEF (Sterner and Wickstrom 1990), and SAFIR (Franssen et al. 2000).

5.3.2.2 Critical Temperatures

Although no standard in the United States specifies critical steel temperatures for performance-based designs, the critical average section temperatures commonly used in the United States are 538°C for steel columns and 593°C for steel beams (similar to temperature acceptance criteria adopted in ASTM E119) regardless of the loads applied to structural members. In Europe (BS 5950 1990, prEN 1993-1-2: 2003, prEN 1994-1-2: 2003, ECCS 2001), critical temperatures for steel members are specified depending on the so-called applied load level, or load ratio, R, i.e., critical temperatures depend on both the type of the structural member and the load level. The critical temperatures, however, are independent of time and also independent of the shape or size of the steel section.

Lawson and Newman (1990) provide a comprehensive summary of critical steel temperatures in accordance with the British Standard (BS 5950 1990), which defines the load ratio as:

$$R = \frac{F}{A_g p_c} + \frac{M_x}{M_b} + \frac{M_y}{p_y Z_y} \qquad \text{for members in compression (columns)} \qquad (5.24a)$$

$$R = \frac{M}{M_c} \qquad \text{for members in bending (beams)} \qquad (5.24b)$$

$$R = \frac{F}{A_e p_y} + \frac{M_x}{M_{cx}} + \frac{M_y}{M_{cy}} \qquad \text{for members in tension} \qquad (5.24c)$$

where:

A_g	=	Gross cross-sectional area (mm^2)
A_e	=	Effective area in tension (mm^2)
p_c	=	Compressive strength of member at room temperature (MPa)
p_y	=	Design strength of steel at room temperature (MPa)
Z_y	=	Elastic section modulus about the minor axis (mm^3)
M_b	=	Buckling resistance moment capacity about major axis at room temperature (N mm)
M_c	=	Bending moment capacity (at room temperature), including the lateral torsional buckling capacity where applicable (N mm)
M_{cx}	=	Bending moment capacity about the major axis at room temperature (N mm)
M_{cy}	=	Bending moment capacity about the minor axis at room temperature (N mm)
F	=	Axial load at the fire limit state (N)
M	=	Applied moment at the fire limit state (N mm)
M_x	=	Applied moment about the major axis at the fire limit state (N mm)
M_y	=	Applied moment about the minor axis at the fire limit state (N mm)

Once the load ratios are calculated from Equations 5.24, the critical temperatures can be determined by interpolating the values in the relevant British Standard (BS 5950 1990) table, reproduced here in Table 5.5. For compression members in braced frames (where the loading is predominantly axial and the bending is comparatively small), the standard distinguishes two ranges of slenderness ratios (cases 1 and 2 in Table 5.5). It should also be noted the standard treats columns in sway frames more conservatively by adopting a single critical temperature of 520°C, regardless of slenderness and load ratios.

For members in bending, BS 5950 (1990) specifies higher critical temperatures for steel beams supporting concrete or composite floors (cases 3 and 4 in Table 5.5) where the beams could be exposed to fire on three sides only. Beams not supporting concrete or composite floors are assigned lower critical temperatures (cases 5 and 6 in Table 5.5). The standard also reduces the critical temperatures for steel and composite beams protected with materials that have not demonstrated their adhesion properties in fire tests (cases 4 and 6 in Table 5.5). Protection materials with proven ability to remain in place during large deformations in fire tests qualify the beams for higher critical temperatures (cases 3 and 5 in Table 5.5).

The critical temperature method could be used in combination with either simplified or advanced heat transfer models to determine the required protection thickness or to confirm that no protection is necessary. Where advanced heat transfer models are used and there is significant temperature variation across the steel section, the hotter flange temperatures should be checked against the critical temperature.

The critical temperature method was developed based on compilations of experimental measurements and analytical extrapolations further confirmed by tests (Lawson and Newman 1990). Very similar concepts were later adopted for the critical temperature methods specified in Eurocode 3 (EN 1993-1-2: 2005), Eurocode 4 (EN 1994-1-2: 2005), and the ECCS Model Code (ECCS 2001), although they resulted in critical temperatures slightly different from those in BS 5950 (1990). Lawson and Newman (1996) discuss these differences in detail.

TABLE 5.5. Critical Temperatures for Structural Steel Members (Lawson and Newman 1990)

Case Number	Description of member	Limiting temperatures (°C) at load ratios:					
		0.7	0.6	0.5	0.4	0.3	0.2
(1) (2)	Braced members in compression: - slenderness ratio ≤ 70 - slenderness ratio ≤ 180	510 460	540 510	580 545	615 590	655 635	710 635
(3) (4)	Members in bending supporting concrete or composite deck floors: - unprotected, or protected with materials with proven "stickability" - all other protected members	590 540	620 585	650 625	680 655	725 700	780 745
(5) (6)	Members in bending not supporting concrete or composite deck floors: - unprotected, or protected with materials with proven "stickability" - all other protected members	520 460	555 510	585 545	620 590	660 635	745 690
(7)	Members in tension	460	510	545	590	635	690

5.3.2.3 Simple Calculation Models

Provisions for simple methods of analysis for the performance-based fire-resistant design of steel and composite structures have been recently introduced in the United States for the first time (AISC 2005). These simple methods are based on AISC design provisions for the capacity of structural steel and composite members at room temperature, adjusted for the deterioration of the mechanical properties of steel and concrete at elevated temperatures. The provisions cover simple methods for tension, compression and flexural steel members and also for composite floor members. The major limitations and simplifications associated with the simple methods of analysis are stated as follows:

"The methods of analysis in this section are applicable for the evaluation of performance of individual members at elevated temperature during exposure to fire. The support and restraint conditions (forces, moments and boundary conditions) applicable at normal temperatures may be assumed to remain unchanged throughout the fire exposure."

Similar simple calculation models have been specified in Europe (BS 5950 1990, prEN 1993-1-2: 2003, prEN 1994-1-2: 2003, ECCS 2001) for more than a decade now, having a similar limitation of applicability to individual members or analysis of parts of the structure only, and using similar assumptions of temperature-independent boundary conditions.

Among simple calculation methods, the most intensive in terms of the calculation effort are probably the methods for the plastic bending moment capacity. The AISC Steel Design Guide 19

(Ruddy et al. 2003) provides a step-by-step procedure for the calculation of flexural capacity of composite floor beams at elevated temperatures, based on earlier research (Ioannides and Mehta 1997) and consistent with newly proposed AISC provisions (AISC 2005). Under fire conditions, the rotational restraint provided by the continuity of the composite floor assembly results in the redistribution of bending moments to beam ends, as illustrated in Figure 5.18, so that the factored applied ultimate bending moment M_u (for the fire load combination) is resisted by both the positive flexural design strength $\phi_b M^{+}{}_n$ and the negative flexural design strength $\phi_b M^{-}{}_n$ of the composite beam, so that:

$$M_u \leq \phi_b \left(M^{+}{}_n + M^{-}{}_n\right) \tag{5.25}$$

Where :

ϕ_b = 0.9 = resistance factor for composite beams in bending
$M^{+}{}_n$ = Positive nominal flexural strength of the composite beam (N mm)
$M^{-}{}_n$ = Negative nominal flexural strength of the composite beam (N mm)

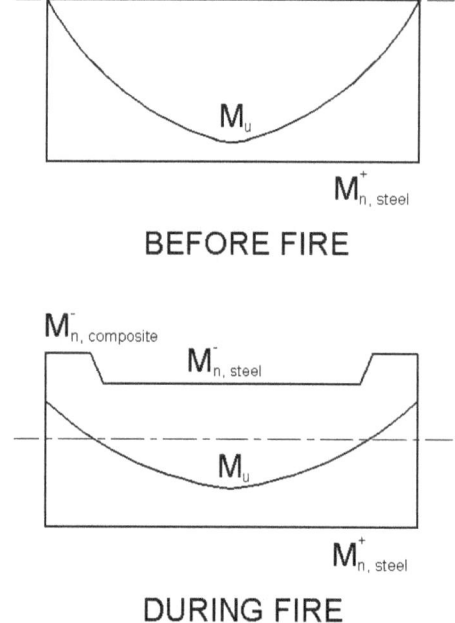

FIGURE 5.18. Moment Envelopes Before and During Fire (Ioannides and Mehta 1997)

The positive nominal flexural strength $M^{+}{}_n$ is determined (in most cases) as the sum of moment couples of the section resultant forces, shown in Figure 5.19. The resultant F_T of tensile forces takes into account the deterioration of steel yield strength in the components of the steel section at their elevated temperatures:

$$F_T = F_{tf} + F_w + F_{bf} \qquad (5.26)$$

where:

F_{tf} = Yield capacity of the top flange at its elevated temperature (N)
F_{tf} = Yield capacity of the web at its elevated temperature (N)
F_{tf} = Yield capacity of the bottom flange at its elevated temperature (N)

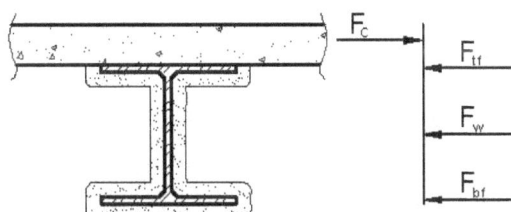

FIGURE 5.19. Typical Positive Moment Force Resultants (Ioannides and Mehta 1997)

The concrete in the compression zone is modeled to achieve its full plastic capacity at the strain of 0.003, assuming the temperature at the top surface of the concrete slab does not increase dramatically. The compressive resultant F_C (equal in magnitude to F_T) would then be located at the middle of the standard equivalent rectangular block of depth a calculated from

$$a = \frac{F_T}{0.85 \, f_c' \, b_f} \qquad (5.27)$$

where:

f_c' = Compressive strength of concrete (Mpa)
b_f = Effective width of concrete slab (mm)

If the depth a of the equivalent compression block is less than the concrete slab depth, then the neutral axis of the composite section lies within the slab, and no further iterations are required. Otherwise, partial compression of the steel section should be assumed, and the location of the neutral axis should be adjusted until the equilibrium of the tensile and compressive resultants in the composite section is satisfied (before the flexural strength is calculated).

The negative nominal flexural strength M_n at the ends of the composite beam is usually provided by the concrete slab reinforcement (parallel with the beam) with the top flange acting together in tension, and forming flexural couples with the remaining portions of the steel beam section acting in compression, as illustrated in Figure 5.20, which assumes the neutral axis passing along the bottom surface of the top flange. The tensile resultant F_T and compressive resultant F_C in this case are:

$$F_T = F_{RB} + F_{tf} \tag{5.28}$$

$$F_C = F_w + F_{bf} \tag{5.29}$$

where:

F_{RB} = Tensile yield capacity of concrete slab reinforcement

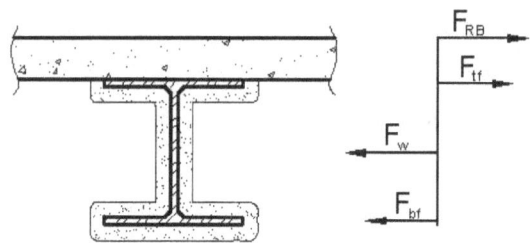

FIGRE 5.20. Typical Negative Moment Force Resultants (Ioannides and Mehta 1997)

If the equilibrium condition for the tensile and compressive resultants is satisfied ($F_T = F_C$), no further iterations are required, and the negative nominal flexural strength is calculated as the sum of the resultant moment couples. Otherwise, the position of the neutral axis should be adjusted until the equilibrium condition is satisfied before the flexural strength is calculated.

For the above procedures (and similar procedures for the plastic bending moment capacity of non-composite steel beams), if the temperatures within the steel or composite section vary considerably, the section could be divided into a larger number of elements so that the temperature in each element would be reasonably approximated as uniform. For a large number of section elements, a suitable computer program is usually used for calculations. One such program, AFCB (available for free download at www.asc.arcelor.com), calculates the capacities of composite steel beams in accordance with Eurocode 4 (EN 1994-1-2: 2005).

5.3.2.4 Advanced Calculation Models

Advanced calculated models could be applied to a single member, an assembly, or to the entire building frame. The newly proposed AISC provisions (AISC 2005) describe the requirement for advanced methods of structural analysis for performance-based fire resistant design as follows:

> "The mechanical response results in forces and deflection in the structural system subjected to the thermal response calculated from the design-basis fire. The mechanical response shall take into account explicitly the deterioration in strength and stiffness with increasing temperature, the effects of thermal expansions and large deformations. Boundary conditions and connection fixity must represent the

proposed structural design. Material properties shall be defined as per Section 4.2.3. The resulting analysis shall consider all relevant limit states, such as excessive deflections, connection fractures, and overall or local buckling."

Similar requirements for advanced calculation models have been specified in Europe (EN 1993-1-2: 2005, EN 1994-1-2: 2005, ECCS 2001) for many years:

- When a global structural analysis for the fire situation is carried out, the relevant failure mode in fire exposure, the temperature-dependent material properties, and member stiffnesses, as well as the effects of thermal expansions and deformations (indirect fire actions), shall be taken into account.

- Advanced calculation models for mechanical response shall be based on the acknowledged principles and assumptions of the theory of structural mechanics, taking into account the effects of temperature. The mechanical response shall also take account of the combined effect of mechanical actions, geometrical imperfections and thermal actions, the temperature-dependent mechanical properties of materials, geometrical nonlinear effects, and the effects of non-linear material properties, including the effects of unloading on the structural stiffness. The effects of the thermally induced strains and stresses, both due to temperature rise and due to temperature differentials, shall be considered. The deformations at ultimate limit state, given by the calculation model, shall be limited as necessary to ensure that compatibility is maintained between all parts of the structure.

- A verification of the calculation results shall be made on basis of relevant test results. The critical results shall be checked, by means of a sensitivity analysis, to ensure that the model complies with sound engineering principles.

To address specific fire resistance problems, several special-purpose finite element computer programs had been developed over the years, most notably SAFIR (Franssen et al. 2000) and VULCAN (Huang et al. 2003). In the last decade, the development of advanced calculation models has intensified significantly in connection with large-scale fire experiments carried out on the Cardington steel-framed building (Kirby 1999), with the main focus on the behavior of unprotected steel-framed composite floors using general-purpose finite element programs, such as ABAQUS, DIANA, LS-DYNA, and ANSYS. The ABAQUS-based model has been reportedly used in practical fire-resistant design applications in the United Kingdom (Ove Arup 2003), and SAFIR was used in at least one practical design in the United States (Chen and Gemeny 2004).

5.4 REFERENCES

TABLE 5.6. Literature references for Figure 5.8

Steels	Reference
WTC 55 ksi and 50 ksi steels	Luecke (2005)
C < 0.3%, C > 0.3%	ASME Boiler and Pressure Vessel Code (2004)
Structural steels	Brockenbrough (1968)
1015 steel	Clark, High-temperature Alloys, Pitman Publishing (1953)
1015 steel	Garafolo, ASTM STP 129 (1952)
Austenite, Ferrite	Koester, Z. Metallk. 39 (1948)
Not specified	Stanzak, Tall (cited in Uddin J Struct Div ASCE 107 [7] p5131 Table 1)

TABLE 5.7. Literature references for Figure 5.11

Steels	Reference
SMA490	Chijiiwa (1993)
ASME-SA516	Gowda, B.C. (1978)
T1, A36	Brockenbrough (1968)
T1a, Cor-Ten, A440, Tri-Ten	Holt (1964)
AS3679.1	Poh (1998)
AS3678 XLERPlate	Chen (2006)
BS4360 Gr 50B	Kirby (1988)
S355, S460M	Outinen (2001)
SM400A	Sakumoto (1999)
60 ksi HSLA plate, 100 ksi plate, A36 12WF161, A36 14WF184, 50ksi A242 angle, A36 angle	Luecke (2005)

Alexander, B. (1982), *Behaviour of Gypsum and Gypsum Products at High Temperatures*, RILEM Committee PHT-44, East Leake, Loughborough, U.K. British Gypsum.

American Institute of Steel Construction (2001), *Manual of Steel Construction, Load and Resistance Factor Design*, 3rd ed., Chicago: American Institute of Steel Construction.

American Institute of Steel Construction (2005), *Specification for Structural Steel Buildings,* ANSI/AISC 360-05, Chicago: American Institute of Steel Construction.

American Society of Civil Engineers (1992), *Structural Fire Protection*, T.T. Lie, ed., ASCE Manuals and Reports on Engineering Practice No. 78, Reston, Va.: American Society of Civil Engineers.

American Society of Civil Engineers (2003), *Standard Calculation Methods for Structural Fire Protection*, ASCE/SEI/SFPE 29-99, Reston, Va.: American Society of Civil Engineers, Structural Engineering Institute, 2003.

Anderberg, Y. (1983), Properties of Materials at High Temperatures-Steel, RILEM Report - Technical Committee 44-PHT, University of Lund, Sweden.

Anderson, C.E., D.E. Ketchum, and W.P. Mountain (1988), *J. of Fire Sciences* 6 (1988) 390-410.

ASME (2004), Boiler and Pressure Vessel Code. Section II Materials, Part D Properties (Customary), Subpart 2 Physical Properties Tables, Table TM-1. ASME International. New York. page 696

Association of the Wall and Ceiling Industries International (1998), *Standard Practice for Testing and Inspection of Field Applied Thin-Film Intumescent Fire-Resistive Materials; an Annotated Guide*, Eaton, R., ed., Technical Manual 12-B, Falls Church, Va.: The Association of the Wall and Ceiling Industries International.

ASTM E21-92 (1992), *Elevated Temperature Tension Tests of Metallic Materials*, West Conshohocken, Pa.: American Society for Testing and Materials.

ASTM C36-97 (1997), *Standard Specification for Gypsum Wallboard*, West Conshohocken, Pa.: American Society for Testing and Materials..

ASTM E119-00 (2000), *Standard Test Methods for Fire Tests of Building Construction and Materials*, West Conshohocken, Pa.: American Society for Testing and Materials.

Baldassarra, C.F., and L.M. Romine (1988), "An Analytical Approach to Fire Protection Engineering," *Building Design and Construction* (March 1988).

Banovic, S. W., C. N. McCowan, and W. E. Luecke. (2005), Federal Building and Fire Safety Investigation of the World Trade Center Disaster: Physical Properties of Structural Steels. NIST NCSTAR 1 3E. National Institute of Standards and Technology. Gaithersburg, MD, September.

Beitel, J., and N. Iwankiw (2002), *Analysis of Needs and Existing Capabilities for Full Scale Fire Resistance Testing*, NIST GCR 02-843, Gaithersburg, Md.: National Institute of Standards and Technology, Building and Fire Research Laboratory.

Brockenbrough, R.L., and B.G. Johnston (1981), *USS Steel Design Manual*, Pittsburgh, Pa.: R.L. Brockenbrough and Assoc. Inc..

BS 5950: Part 8, *Structural Use of Steelwork in Buildings, Code of Practice for Fire Resistant Design* (1990), U.K.: British Standards Institution.

Butler, K.M. (1997), "Physical Modeling of Intumescent Fire Retardant Polymers," Chapter 15 of *Polymeric Foams Science and Technology*, ASC Symposium Series 669, Washington: American Chemical Society, pp. 214-230.

Carino, N.J., M.A. Starnes, J.L. Gross, J.C. Yang, S. Kukuck, K.R. Prasad, and R.W. Bukowski, (2005), Federal Building and Fire Safety Investigation of the World Trade Center Disaster: Passive Fire Protection. NIST NCSTAR 1-6A. National Institute of Standards and Technology. Gaithersburg, MD, September

Chen, J., B. Young, and B. Uy (2006), Behavior of high strength structural steel at elevated temperatures. J. Struct. Eng. ASCE, 132(12):1948–1954.

Chen, F.F., and D.F. Gemeny (2004), "Case Study Using SAFIR to Predict Fire Resistance of a Column in a Performance Based Environment," SFPE Fire Protection Engineering Magazine 23 (Summer 2004) 29-35.

Chijiiwa, R., H. Tamehrio, Y. Yoshida, K. Funato, T. Uemori, and Y. Horii (1993), Development and practical application of fire-resistant steel for buildings. Nippon Steel Technical Report 58, Nippon Steel, 1993. UDC669.14.018.291 : 699.81.

Clark, C.L. 1953. High-temperature Alloys. Pitman, New York.

Cooke, G.M.E. (1988), "An Introduction to the Mechanical Properties of Structural Steel at Elevated Temperatures," *Fire Safety Journal* 13 (1988) 45-54.
Design of Fire Resistive Assemblies with Steel Joists (Design 2003), Technical Digest 10, Myrtle Beach, S.C.: Steel Joist Institute.

Directory of Listed Building Products, Materials and Assemblies (OPL 2004), Elmendorf, Tex.: Omega Point Laboratories Inc.

ECCS Technical Committee 3 (1983), *European Recommendations for the Fire Safety of Steel Structures, Calculation of the Fire Resistance of Load Bearing Elements and Structural Assemblies Exposed to the Standard Fire*, Amsterdam: Elsevier Scientific Publishing Company.

ECCS Technical Committee 3 (1995), *Fire Resistance of Steel Structures*, ECCS Publication No. 89, Brussels: European Convention for Constructional Steelwork.

ECCS Technical Committee 3 (2001), *Model Code on Fire Engineering*, Brussels: European Convention for Constructional Steelwork.

EN 1993-1-2: 2005, *Eurocode 3: Design of Steel Structures—Part 1-2: General Rules, Structural Fire Design*, Stage 49 Draft, Brussels: European Committee for Standardisation, 2005.

EN 1994-1-2: 2003, *Eurocode 4: Design of Composite Steel and Concrete Structures—Part 1-2: General Rules, Structural Fire Design*, Stage 34 Draft, Brussels: European Committee for Standardisation, 2003.

ENV 13381-1: 2005, *Test Methods for Determining the Contribution to the Fire Resistance of Structural Members—Part 1: Horizontal Protective Membranes*, Brussels: European Committee for Standardisation, 2005.

ENV 13381-2: 2002, *Test Methods for Determining the Contribution to the Fire Resistance of Structural Members—Part 1: Vertical Protective Membranes*, Brussels: European Committee for Standardisation, 2002.

ENV 13381-4: 2002, *Test Methods for Determining the Contribution to the Fire Resistance of Structural Members—Part 1: Applied Protection to Steel Members*, Brussels: European Committee for Standardisation, 2002.

Franssen, J.M., V.K.R. Kodur, and J. Mason (2000), *User's manual for SAFIR2001: a computer program for analysis of structures submitted to the fire*, Liege, Belgium: University of Liege.

Garofalo, F., Malenock, P.R., and Smith, G.V. (1952), The Influence of Temperature on the Elastic Constants of Some Commercial Steels. Symposium on Determination of Elastic Constants. ASTM Special Technical Publication 129. 10-27.

Gewain, R.G., E.W.J. Troup (2001), "Restrained Fire Resistance Ratings in Structural Steel Buildings," *Engineering Journal* 38:2 (2001).

Gewain, R.G., N.R. Iwankiw, and F. Alfawakhiri (2003), *Facts for Steel Buildings: Fire*, Chicago: American Institute of Steel Construction.

Gowda, B.C. (1978), Tensile properties of SA516, grade 55 steel in the temperature range of 25 °C–927 °C and strain rate range of 10^{-4} to 10^{-1} sec^{-1}/in. George V. Smith, editor, Characterization of Materials for Service at Elevated Temperatures, pages 145–158, New York, 1978. The American Society of Mechanical Engineers. Presented at the 1978 ASME/CSME Montreal Pressure Vessel & Piping Conference, Montreal, Quebec, Canada, June 25–29 1978.

Gypsum Association (1998), *Gypsum Board: Typical Mechanical and Physical Properties*, GA-235-98, Washington: Gypsum Association.

Hamins, A. (1998), *Evaluation of Intumescent Body Panel Coatings in Simulated Post Accident Vehicle Fires*, NISTIR 6157, Gaithersburg, Md.: National Institute of Standards and Technology, Building and Fire Research Laboratory.

Harmathy, T.Z. (1983), *Properties of Building Materials at Elevated Temperatures*, DBR Paper No. 1080, Ottawa: National Research Council of Canada, Division of Building Research.

Harmathy, T.Z., and W.W. Stanzak (1970), "Elevated Temperature Tensile and Creep Properties of Some Structural and Prestressing Steels", *ASTM STP 464*, West Conshohocken, Pa.: American Society for Testing and Materials, pp 186-208.

Holt, J. M. (1964) Short-time elevated-temperature tensile properties of USS Cor-Ten and USS Tri-Ten high-strength low-alloy steels, USS Man-Ten (A 440, high strength steel, and ASTM A 36 steel. Progress Report 57.19-901(1), United States Steel Corporation.

Huang, Z., I.W. Burgess, and R.J. Plank (2003), "Modelling Membrane Action of Concrete Slabs in Composite Buildings in Fire. Part I: Theoretical Development," *ASCE Journal of Structural Engineering* 129:8 (2003) 1093-1102.

Huang, Z., I.W. Burgess, and R.J. Plank (2003), "Modelling Membrane Action of Concrete Slabs in Composite Buildings in Fire. Part II: Validations," *ASCE Journal of Structural Engineering* 129:8 (2003) 1103-1112.

International Building Code (ICC 2000), Falls Church, Va.: International Code Council.

Intertek Testing Services NA Inc. (2004), *Directory of Listed Products* (ITS 2004), Cortland, N.Y.: Intertek Testing Services NA Inc.

Ioannides, S.A., and S. Mehta (1997), "Restrained vs. Unrestrained Fire Ratings: A Practical Approach," *Modern Steel Construction* (May 1997).

Jerath, V., K.J. Cole, and C.I. Smith (1980), *Elevated Temperature Tensile Properties of Structural Steels Manufactured by the British Steel Corporation*, Report T/RS/1189/11/80/C, BSC Teesside Laboratories (July 1980).

Kirby, B. (1983), *The Behaviour of Structural Steels Manufactured by BSC under Stress Controlled Anisothermal Creep Conditions*, Report SH/RS/3664/4/83/B, BSC Sheffield Laboratories (October 1983).

Kirby, B. (1999), *The Behaviour of Multi-Storey Steel Framed Buildings in Fires, A European Joint Research Program*, U.K.: British Steel Swinden Technology Centre.

Kirby, B.R., and Preston (1988), "High Temperature Properties of Hot-Rolled, Structural Steels for Use in Fire Engineering Design Studies," *Fire Safety J.* 13 (1988) 27-37.

Kodur, V.K.R., and T.Z. Harmathy (2002), "Properties of Building Materials," Chapter 10 of *The SFPE Handbook of Fire Protection Engineering*, 3rd ed., Bethesda, Md.: Society of Fire Protection Engineers.

Köster, W. (1948), Die Temperaturabhängigkeit des Elastizitätsmoduls reiner Metalle, Z. Metallkd. 39(1).1-9.

Lane, B. (2000), "Performance-Based Design for Fire Resistance," *Modern Steel Construction* (December 2000).

Lawson, R.M., and G.M. Newman (1990), *Fire Resistant Design of Steel Structures—A Handbook to BS5950:Part 8*, SCI Publication 080, U.K.: The Steel Construction Institute.

Lawson, R.M., and G.M. Newman (1996), *Structural Fire Resistant Design to EC3 & EC4, and Comparison with BS 5950*, SCI Publication 080, U.K.: The Steel Construction Institute.

Lawson, R.M., D. Oshatogbe, and G.M. Newman (2002), *Design of Fabsec Beams in Non-Composite and Composite Applications (including Fire)*, U.K.: Fabsec Limited.

Luecke, W.E., J.D. McColskey, C.N. McCowan, S.W. Banovic, R.J. Fields, T. Foecke, T.A. Siewert, and F.W. Gayle (2005), Federal building and fire safety investigation of the world trade center disaster: Mechanical properties of structural steel. Technical Report NCSTAR 1- 3D, National Institute of Standards and Technology. Available on-line at http://wtc.nist.gov.

Lund Institute of Technology (1983), *Properties of Materials at High Temperatures: Steel*, Anderberg, Y., ed., Report LUTVDG/TVBB-3008), Lund: Sweden, Lund Institute of Technology.

Malhotra, H.L. (1982), *Design of Fire-Resisting Structures*, New York: Chapman and Hall.

NAHB Research Center (2004), *Residential Steel Framing: Builder's Guide to Fire and Acoustic Details*, U.S. Department of Housing and Urban Development (HUD) (June 2004).

NFPA 251 (1999), *Standard Methods of Tests of Fire Endurance of Building Construction and Materials*, Quincy, Mass.: National Fire Protection Association.

NFPA 5000 (2003), *Building Construction and Safety Code*, Quincy, Mass.: National Fire Protection Association.

Outinen, J., O. Kaitila, and P. Mäkeläinen (2001), High-temperature testing of structural steel and modelling of structures at fire temperatures. Report TKKTER- 23, Helsinki University of Technology, Laboratory of Steel Structures.

Ove Arup & Partners Massachusetts Inc. (2003), "Task 1: Assessment of Technical Issues and Needs," *Strategy for Integrating Structural and Fire Engineering of Steel Structures*, Report for the American Institute of Steel Construction.

Pelanne, C.M. (1978), "Does the Insulation Have a Thermal Conductivity? The Revised ASTM Test Standards Require an Answer," *Thermal Transmission Measurements of Insulation,* ASTM STP 660, West Conshohocken, Pa.: American Society for Testing and Materials, 60-70.

Poh, K.W. (1998), Behaviour of Load-Bearing Members in Fire. PhD thesis, Dept. of Civil Engineering, Monash University, Clayton, Victoria, Australia.

Ruddy, J., J.P. Marlo, S.A. Ioannides, and F. Alfawakhiri (2003), *Fire Resistance of Structural Steel Framing*, AISC Steel Design Guide 19, Chicago: American Institute of Steel Construction.

Sakumoto, Y. (1999), Research on new fire-protection materials and fire-safe design. J. Struct. Engineering, ASCE, 125:1415–1422.

SFPE Task Group on Documentation of Fire Models (1995), *A Practical User's Guide to Fires—T3, A Three-Dimensional Heat-Transfer Model Applicable to Analyzing Heat Transfer Through Fire Barriers and Structural Elements*, Bethesda, Md.: Society of Fire Protection Engineers.

Siu, J.C. (2005), "Challenges Facing Engineered Structural Fire Safety—A Code Official's Perspective," *SFPE Fire Protection Engineering Magazine* 25 (Winter 2005) 30-36.

Sterner, E., and U. Wickstrom (1990), *TASEF—Temperature Analysis of Structures Exposed to Fire,* Fire Technology Report 1990:05, Boras, Sweden: Swedish National Testing Institute.

Sultan, M.A. (1996), "A Model for Predicting Heat Transfer Through Noninsulated Unloaded Steel-Stud Gypsum Board Wall Assemblies Exposed to Fire," *Fire Technology* 32:3 (1996) 239-259.

Touloukian, Y.S., Kirby, R.K., Taylor, R.E. and Klemens, P.D. (1977), Thermal Expansion Metallic Elements and Alloys, IFI/Plenem, New Your, NY.

Uddin T. and C. G. Culver (1975), Effects of Elevated Temperature on Structural Members. J. Struct. Div.ASCE. 101 (7). 1531-1549.

UL 263 (2003), *Fire Tests of Building Construction and Materials*, 13th ed., Northbrook, Ill.: Underwriters Laboratories Inc.

Underwriters Laboratories Inc. (2004), *Fire Resistance Directory*, Vol. 1, Northbrook, Ill.: Underwriters Laboratories Inc.

Yandzio, E., J.J. Dowling, and G.M. Newman (1996), *Structural Fire Design: Off-Site Applied Thin Film Intumescent Coatings*, SCI Publication 160, U.K.: The Steel Construction Institute.

Chapter 6

General Application of Guidelines

Donald O. Dusenberry, P.E., Simpson Gumpertz & Heger Inc.
Morgan J. Hurley, P.E., Society of Fire Protection Engineers

Performance-based design of structures for fire resistance requires the collaboration of fire protection engineers and structural engineers. Fire protection engineers typically analyze fire environments and the transfer of heat from the fire to the elements of a structure. Structural engineers use the results of the heat transfer analysis to determine the structural response, considering thermally induced strains and the effects of changes of material properties at elevated temperatures. Typically this collaboration is most effective when fire resistance is considered as early as the conceptual design stage, when flexibility in structural concepts allows the fire resistance considerations to influence the structural design.

Designing structural fire resistance on a performance basis generally includes the following steps (see SFPE 2006, SFPE 2007 and Aktan 2007):

1. Predict the thermal environment surrounding a building structure (or portion thereof) in the event of a fire. This is generally accomplished by modeling the fire.
2. Determine the thermal response of the structure by conducting heat transfer analyses. Most fire models provide temperature boundary conditions, which must be translated into heat flux boundary conditions to determine the thermal response of the structure.
3. Evaluate the structural response. This involves evaluation of the impacts of temperature-dependent material properties and stresses due to thermal expansion and contraction. Hence, material and geometry nonlinearities often are important.

Underlying these three steps, if a risk assessment is part of the design, is a risk analysis for the identified fire hazards, the response of the structural system, and the associated consequences of failures that might be induced by fires.

This chapter provides general guidance on the approaches to and practical aspects of implementing a fire resistant design for conventional applications.

Section 6.1 reviews key concepts for identifying performance objectives and conducting risk analyses, with examples. Section 6.2 discusses considerations for identifying fire scenarios and developing fire exposure curves. Sections 6.3 and 6.4 discuss heat transfer analyses and analyses of structural response to elevated temperatures. Section 6.5 discusses fire resistant structural design. Sections 6.6 and 6.7 discuss design issues particular to concrete and steel structures, respectively.

6.1 PERFORMANCE OBJECTIVES AND RISK ASSESSMENT

Chapter 3 presented a detailed framework for developing performance goals and objectives, conducting risk assessments, and selecting a risk mitigation approach for performance based design of fire resistant conventional construction. This section discusses implementing this approach in the design process.

6.1.1 Performance Objectives

Performance objectives, which must be fulfilled to achieve the goals of the performance-based design, often are derived from code-based requirements, sometimes with additional goals that satisfy the mission of the owner or other parties. For instance, a primary goal in building codes is life safety of building occupants. Other goals might address protection of property, business continuity, or life safety risk for firefighters.

Specific criteria should be established for each performance objective to identify how it will be met. For instance, structural performance objectives for life safety could provide structural resistance that allows adequate time for occupant evacuation (note that this does not address firefighter safety or preservation of building contents).

Overall fire safety design strategies for a building might include:

- Appropriate compartmentation for limiting heat transmission and fire and smoke spread
- Intact egress paths or safe havens
- Appropriate passive fire protection for the structural system
- Active fire suppression systems to control fire growth

A subset of fire safety design strategies for the structural system might include:

- Structural integrity to ensure that architectural features remain intact (e.g., limit frame deflections so that fire barriers are not breached)
- Adequate structural stiffness and strength for columns, floors, and connections

Other performance objectives for fire resistant design might include life safety of firefighters (e.g., no partial or global collapse of the structure), protection of property (including adjacent buildings), survival of building contents (e.g., bank, library, or museum contents), protection of infrastructure (e.g., integrity of gas lines or electrical substations), or protection of the environment (e.g., no chemical spills).

6.1.2 Risk Assessments

The risk analysis should be addressed at the start of the project, and continued through the design process. Owners and their consultants must make decisions to manage and mitigate risk through combinations of prudent design and risk acceptance.

For a risk assessment to be useful, it must be expressed in quantitative terms. Risk can be defined as the potential rate at which consequences (e.g., economic, life, or other types of losses) may occur for a given structure and its associated hazards.

Chapter 3 states that the probability of a loss can be evaluated from the following equation:

$$P[Loss > \theta\,] = \Sigma_H \Sigma_{LS} \Sigma_D\ P[Loss > \theta | D]\ P[D|LS]\ P[LS|H]\ P[H] \qquad (3.1)$$

Where:

P[A]	=	Probability of event A
P[A|B]	=	(Conditional) probability of event A, given the occurrence of event B
θ	=	An appropriate loss metric: severe injury or death, direct damage costs, loss of opportunity costs, etc.
P[LS|H]	=	Conditional probability of a structural limit state
P[D|LS]	=	Conditional probability of damage state (e.g., negligible, minor, moderate, severe)
P[H]	=	Probability of a hazard occurrence
P[Loss > θ|D]	=	Conditional probability of loss

As an alternative, the risk assessment may be based on a set of stipulated scenario events rather than on a hazard with a random intensity, depending on the preferences of the decision-maker. For a fire scenario, the risk metric in Eq. 3.1 becomes a conditional probability:

$$P[Loss > \theta | H_s] = \Sigma_{LS} \Sigma_D\ P[Loss > \theta | D]\ P[D|LS]\ P[LS|H_s] \qquad (3.2)$$

where:

H_s = Fire scenario event(s)

These expressions from Chapter 3 can be simply restated as:

$$Risk = \Sigma_H\ Probability\ of\ failure\ x\ Consequence\ of\ failure$$

where the summation is evaluated for the fire hazard, H, or for all identified fire scenarios, H_s.

For a risk analysis, the fire hazard can be expressed either probabilistically or as a set of fire scenarios (see Ch. 3) with a probability of occurrence for each fire scenario. While tables of fire occurrence rates are useful, the designer must be careful to include other factors that may affect an individual building. For instance, buildings sited in locations where wildfires occur could have a greater probability of fire ignition than buildings that are remote from such hazards. Occupancies that involve storage of volatile fluids and buildings with outdated and overloaded electrical systems could have a greater probability of ignition.

Unfortunately, there might not be reliable data leading to precise assessments of fire hazards under unusual conditions. In some cases, data might be held by organizations that monitor fire occurrences (e.g., the National Fire Protection Association). In many cases the engineer must apply judgment to determine factors for fire hazards that will contribute to the risk assessment.

Fire suppression systems affect the frequency of fully developed fires, given that fire ignition has occurred. The effect of these systems generally is considered only in risk and probabilistic analyses of fire occurrence rates (see Section 3.2.2). In scenario-based analyses, effectiveness and reliability of suppression systems are considered only when selecting fire scenarios to be evaluated.

The probability of failure (i.e., exceeding a structural limit state or failing a performance objective), and the resulting damage state, are based on the results of structural analyses. Since structural behavior can be highly nonlinear when thermal effects and temperature dependent properties of construction materials are included in the analysis, care must be given to analysis approaches (e.g., linear approaches usually are insufficient) and a clear definition of what constitutes "failure" must be established in consideration of the selected analysis approach.

Risk analysis can be used to make comparisons against risks for similar types of construction or to identify the factors that contribute to the overall risk. A comparative risk analysis cannot guarantee specific levels of performance, but it can provide guidance on how to (1) reduce risks associated with the performance objectives, and (2) develop prudent and effective designs based on the best available information.

6.2 DESIGN FIRES

Fire protection engineers, through collaboration with structural engineers, must identify the relevant fire environments for structural analyses and design. This requires consideration of the factors that affect fire severity, fire growth within compartments, design fire definitions, fuel loads, fire resistance ratings, and fire modeling.

6.2.1 Factors That Affect Fire Severity

Each fire and its severity are affected by the following factors:

- The arrangement of fuel in the fire compartment
- The composition and amount of fuel
- Ventilation characteristics
- The geometry of the compartment
- Thermal properties of the compartment walls, floor and ceiling

6.2.1.1 Fuel Arrangement

The arrangement of fuel determines whether a fire can be treated as uniform over a compartment or as a discrete fire location within a compartment. If the fuel is well distributed over the floor area of the compartment, it is usually assumed that the fire conditions are uniform throughout the enclosure. Conversely, if the fuel is concentrated over a portion of a floor area, an assumption that the fire exposure would not be uniform would be more appropriate.

6.2.1.2 Composition and Amount of Fuel

The composition and amount of fuel in an enclosure can affect the peak temperatures and the duration of burning. In Figure 6.1, characteristic curves plotted using the method developed by

174

Magnusson and Thelandersson (1970) show the theoretical relationship between peak fire temperature and time as a function of fuel load in the compartment.

All predictive methods were developed on the basis of wood (or other cellulosic materials) as fuel (SFPE, 2004). Some methods specify the fuel load on the basis of mass, while others require input on the basis of caloric value of fuel. The methods developed based on wood crib fires reasonably predict maximum fire temperatures. However, special attention is required when compartments contain significant quantities of non-cellulosic fuels and ventilation is not expected to control the burning rate. Predictive methods based on burning wood cribs should be appropriate for most compartments of interest.

Many hydrocarbon-based materials, such as plastics, have approximately twice the heat of combustion of wood. More energy is required to cause wood to release flammable vapors than is needed for plastics (Karlsson and Quintiere 1999), so that wood has a lower rate of mass loss and longer burning durations. However, heat energy that is released per mass of available oxygen is essentially constant for a wide variety of fuels (Huggett 1980; Drysdale 1999).

Two approaches can be used to estimate the fuel load in a compartment. Typically for design, fuel loads can be determined from tabulated values that are based on occupancy. Alternatively, for an existing structure or for occupancies that are not included in tables of fuel loads, combustible furnishings and contents can be inventoried. Masses, and therefore fuel loads, of the combustible items anticipated for the compartment can be determined by their weights or by measuring their dimensions and multiplying their volumes by effective densities.

Chapter 3 summarized available information on expected fire loads (fuels) in buildings with various occupancies. Tables 3.3(a), 3.3(b), and 3.3(c) relate results from different studies conducted at different times over a period of approximately 25 years. When assessing fire loads, engineers need to consider matters such as whether the expected occupancy of the building under design would generate fire loads that are atypical for the general occupancy category and building type.

It is generally acceptable to rely on published data. However, there are circumstances for which specific evaluations of fire ignition risks and fire loads are important. When a specific building will house activities that involve conditions not normally found in similar occupancies, specific evaluations can be essential. An example could be an office building that will contain fuel storage capacity for backup site power generation. The presence of petroleum fuel storage and handling capability in this application warrants special consideration. In this example, the engineer should assess:

- The quantity of fuel on the premises
- The reliability of the fuel containment systems
- The fire load generated by a spill (SFPE 2002, pp. 3-24 to 3-26) relative to fire loads normally associated with office building occupancies

6.2.1.3 Ventilation Characteristics and Compartment Geometry

Compartment gas temperatures and burning rates are influenced by the available ventilation. In ventilation-limited fires, the rate of airflow into the enclosure will govern the heat release rate inside the enclosure, and fuel vapors that cannot burn inside the enclosure will burn outside once they encounter fresh air.

Fuel in well-ventilated compartments (e.g., with ventilation factor $F_v = 0.12$ m$^{-1/2}$ in Figure 6.2) tends to burn more quickly and create higher peak temperatures than fuel in compartments that have restricted airflow (e.g., with $F_v = 0.02$ m$^{-1/2}$ in Figure 6.2). Hence, access to oxygen during a fire must be evaluated.

The shapes and locations of openings in a fire compartment are important. Most fire exposure curves assume openings in walls (rather than floors or ceilings) and that the associated airflow is induced by buoyancy. When airflow is buoyancy driven, the flow rate through openings is a function of the distance above or below the plane at which the pressure in a compartment is the same as the pressure outside the compartment (the neutral plane) (Drysdale 1999). Above the neutral plane, the pressure is higher than the ambient pressure, and heated gas flows out of the compartment through openings. Below the neutral plane the pressure is lower than ambient, so air flows into the compartment.

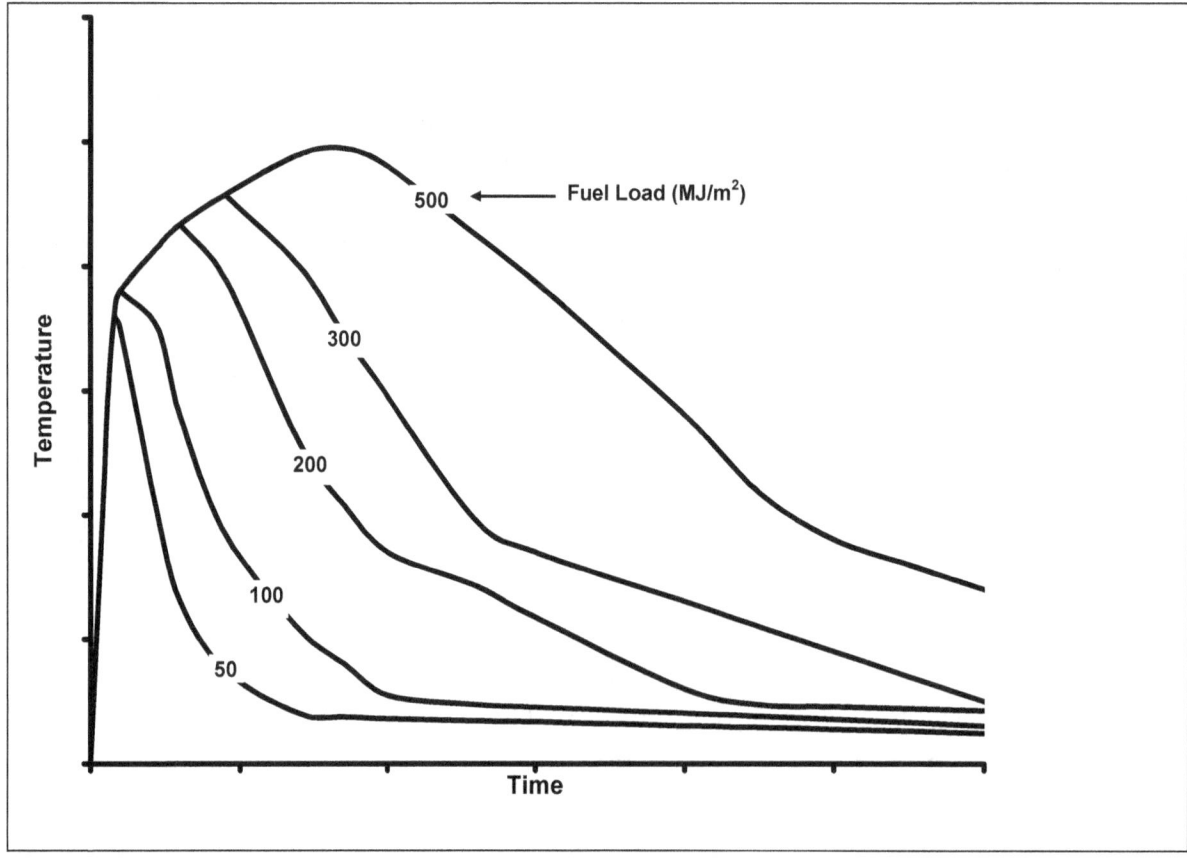

FIGURE 6.1. Relationship between Temperature and Time as a Function of Fuel Load (adapted from Magnusson and Thelandersson, 1970)

Given ventilation openings in vertical surfaces, tall window openings tend to ventilate compartments more efficiently than do shallow windows, resulting in a higher ventilation factor and possibly hotter and shorter fires. Through tall windows, convection encourages cool air with oxygen to flow into the compartment in the lower portion of the window and hot fumes generated by the fire to flow out the upper portion of the window. Convection currents are weaker when windows are shallow in height. This fact is represented in the ventilation factor, F_v, used in the Swedish curves (Magnesson and Thelandersson 1970).

Fire exposure curves, such as those shown in Fig. 6.1, and empirical equations are generally based on an assumption that ventilation openings are in vertical surfaces. Compartments with openings in horizontal surfaces, such as stairwell openings between floors in office spaces, require special attention with methods applicable to such conditions (see Section 6.2.5.3).

Predictive methods based on available fire exposure curves are also based on an assumption that the compartments are "well stirred," i.e., that the temperature in the compartment is uniform. This assumption can break down in compartments that are irregularly shaped. SFPE (2004) investigated the ability of various methods to predict fires in compartments that were long and narrow with the ventilation opening on one of the shorter sides. This study showed that some existing methods provide reasonable predictions of fire severity in long and narrow compartments. See Section 6.2.2 for additional information.

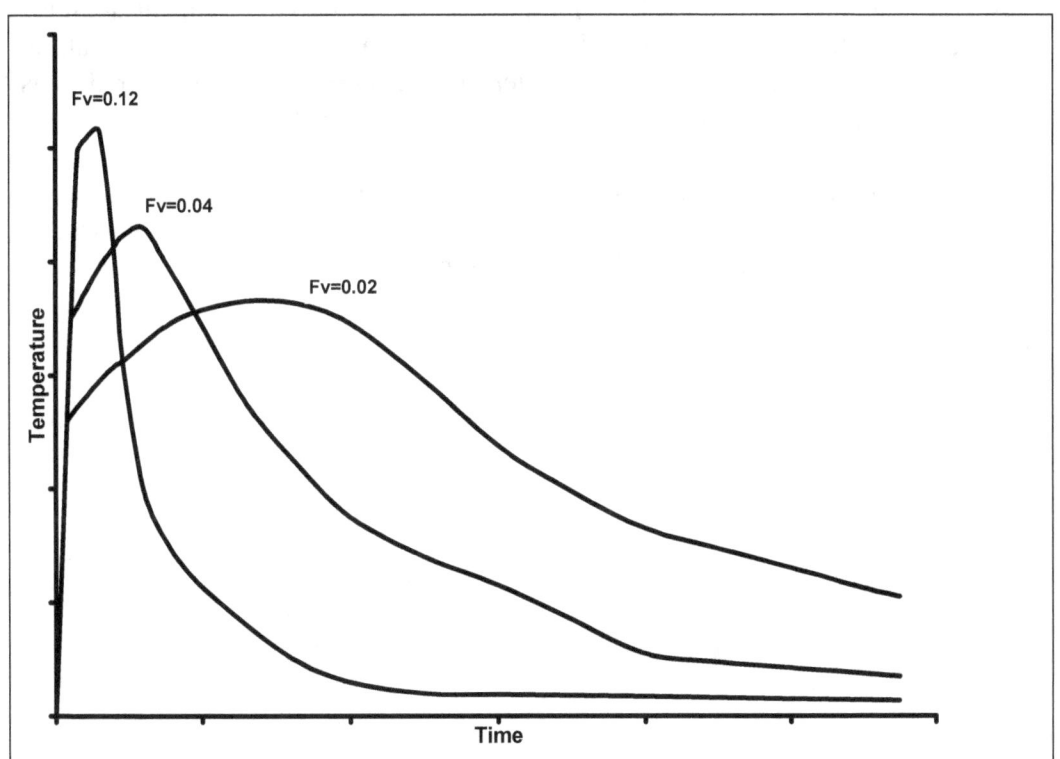

FIGURE 6.2. Relationship between Fire Temperature and Time as a Function of Ventilation (adapted from Magnusson and Thelandersson, 1970)

6.2.1.4 Thermal Properties

The thermal properties of the construction materials and furnishings in an enclosure also influence gas temperature histories during fires. Gas temperatures in compartments with low thermal inertia (defined as the product of density, specific heat, and thermal conductivity) tend to be hotter than gas temperatures in compartments with higher thermal inertias (SFPE 2004). The surfaces with higher thermal inertia absorb more heat energy from the fire-induced environment, reducing gas temperatures, whereas less heat energy is absorbed by materials with lower thermal inertia.

6.2.2 Fires in a Compartment

The rate of burning in fully developed fires is primarily controlled by either ventilation or fuel. In ventilation-controlled burning, the rate of combustion depends on the volume of available air; if there is insufficient air, combustibles will be incompletely burned and flames may extend out the windows to complete combustion of the unburned gaseous fuel mixture. In fuel-controlled burning, the rate of combustion depends on the surface area of the fuel, especially in large well-ventilated rooms with limited amounts of combustible surfaces.

As fires develop in compartments, the increasing gas temperatures in the upper layer of the compartment generate increasing radiant heat flux that heats all objects in the compartment. At a critical level of heat flux, all exposed combustible items in the compartment simultaneously begin to burn, leading to a rapid increase in both the heat release rate and gas temperatures. This transition is referred to as flashover, and fires after this transition are often referred to as 'post-flashover' fires or 'fully developed' fires, as shown in Fig. 6.3 (Buchanan 2001).

The primary threat to most structures arises during the fully-developed stage of burning and, for insulated steel or concrete structures, possibly during the decay stage as cooling occurs (SFPE 2004). Figure 6.3 illustrates the phases of fire development for a room fire.

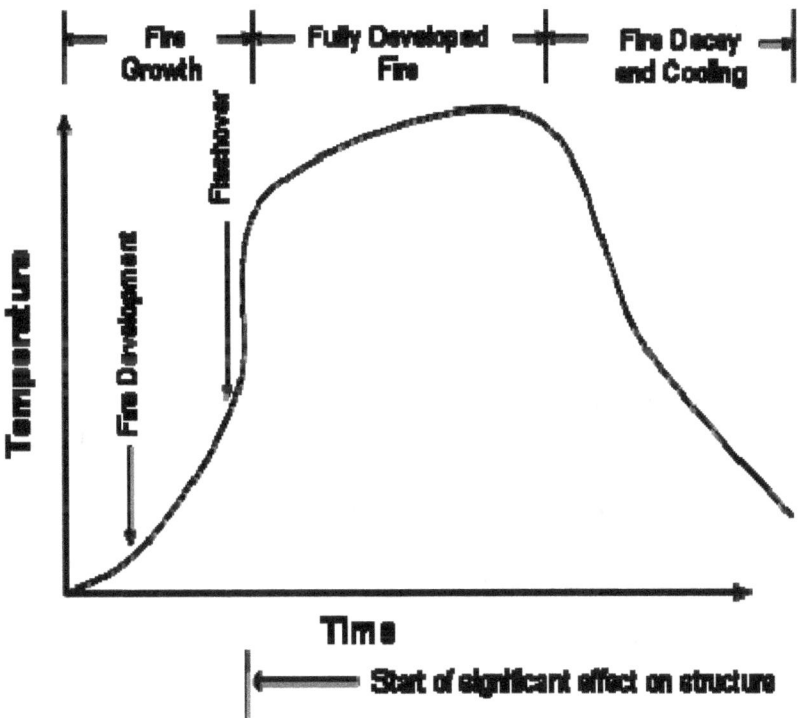

FIGURE 6.3. Phases of Fire Development for a Fire in an Enclosure (adapted from SFPE 2004).

Generally, fully developed fires are assumed to involve entire rooms or enclosures. Where fire barriers are not provided, the entire building, or at least an entire floor, generally is assumed to be fully involved.

If a building contains multiple areas that are enclosed with fire-rated boundaries, then it may be necessary to consider separate fire scenarios in each enclosed area. Fire rated boundaries include partitions with fire-rated door and wall construction that extends from floor slab to floor slab or from the floor slab to a fire-rated suspended ceiling. Partitions that stop at unrated suspended ceilings are not fire boundaries.

6.2.3 Design Fires and Fire Scenarios

A design fire is defined to be a time-temperature curve of the gases in one or more compartments and possibly on multiple floors and a fire scenario describes all the conditions required for a particular fire event. A time-temperature curve, or fire exposure curve, is the quantitative measure of the fire scenario and its supporting conditions.

Factors that affect fire development include the following conditions:

- Compartment size
- Furnishings and contents
- Fuel properties
- Ignition sources

179

- Ventilation conditions
- First item ignited and its location

A large number of design fire scenarios may be possible. A set of credible fire scenarios should be selected to represent the fire hazard. Resources from SFPE (2007), ICC (2006), and NFPA (2006) can provide guidance, information, and methodologies for developing fire design scenarios.

6.2.4 Predictive Methods for Fire Modeling

There are three approaches to generating time-temperature data for analysis:
1. Time equivalence methods that use standard fire data in combination with compartment features.
2. Fire exposure curves and/or parametric equations based on experimental data of realistic fire conditions, as described in Section 2.5.2 and shown in Fig. 3.2.
3. Computer analysis of fire scenarios

6.2.4.1 Time Equivalence Methods

Time equivalence methods allow designers to use standard fire exposures as a basis for qualifying fire protection products and systems under prescriptive building code approaches, as discussed in Section 2.5.2. Time equivalents are determined by adjusting time-temperature data for compartment fires to determine the duration in the standard test that would have the same heating effect as a compartment fire.

As with the standard fire test, time equivalent methods only consider individual elements or subsystems; structural system behavior cannot be addressed with this approach. Time equivalence methods are suitable for compartments with similar dimensions for width and length; long, narrow compartments are not satisfactorily correlated by these methods (Law 1997).

6.2.4.2 Fire Exposure Curves and Parametric Equations

For most performance-based designs, fire exposure curves or parametric equations are satisfactory approximations of actual fire conditions that occur in a building. Most fire exposure curves and parametric equations approximate temperatures in fully developed fires as uniform throughout the compartment (the one-zone model). Buchanan (2001) and SFPE (2004) describe the underlying assumptions for and applicability of many of the available fire exposure curves.

Fire exposure curves and equations have been developed to account for the effects of the following parameters:

- Fuel load (unit weight)
- Area and height of ventilation areas
- Enclosure geometry (width, depth, height, and total surface area of enclosure)
- Thermal properties of compartment construction materials and furnishings

The amount of ventilation in a compartment is often described by the ventilation factor (or opening factor) F_v:

$$F_V = \frac{A_V \sqrt{H_V}}{A_t} \quad (\text{m}^{-1/2}) \tag{6.1}$$

Where:

       A_V = Area of openings
       A_t = Total surface area of the compartment
       H_V = Height of openings

In some compartments, the assumption of uniform temperatures does not hold. Compartment geometry and the distribution of ventilation sources can cause the fire to burn preferentially close to sources of ventilation, and for the fire to move away from the ventilation opening(s) only as the fuel is consumed. Nevertheless, some models can reasonably predict fire conditions in compartments where preferential burning occurs (SFPE 2004, McGrattan 2002).

Compartment fires where the ventilation factor changes over time are difficult to analyze using available curves and equations. None of these references accounts directly for window breakage or partition failure. Designers using these references often overcome this limitation by selecting ventilation characteristics that yield an upper bound fire exposure. This process may involve investigating several realistic estimates of ventilation factors, since both peak temperature and fire duration are affected by ventilation.

Law's method (1983), which estimates the maximum gas temperature, is appropriate for all roughly cubic compartments (height to width ratio between 0.5 and 2.0) and for long, narrow compartments with a ventilation factor, F_v, greater than 0.56 $\text{m}^{\frac{1}{2}}$ created by openings on only one of the shorter sides.

Law's method does not predict temperatures during the decay phase. To account for the fire decay phase while using Law's method, designers often assume a decay rate of 7 °C/min for fires with predicted durations of 60 minutes or more, or 10 °C/min if the predicted duration is less than 60 min.

The Magnusson and Thelandersson method and the Lie method can be used for long, narrow, under-ventilated compartments with openings on one of the shorter sides. The Magnusson and Thelandersson method is appropriate for $1/F_v$ in the range of 45 $\text{m}^{-\frac{1}{2}}$ to 85 $\text{m}^{-\frac{1}{2}}$, the Lie method is appropriate when $1/F_v$ is approximately 345 $\text{m}^{-\frac{1}{2}}$.

6.2.4.3 Fire Modeling with Computer Programs

Three classes of fire models have been programmed for computer analyses of gas temperatures in a room or compartment..

1. Empirical models are based on correlations with full-scale fire experiments.

2. "Zone" models treat the compartment as either one or two homogeneous volumes in which conservation of mass, momentum, and energy are computed in a dynamic process.
3. "Field" models are based on computational fluid dynamics (CFD). Field models use a three-dimensional grid to model the changes in gas and surface temperatures over the room volume.

Some of these models analyze for a user-specified heat release rate; some predict the heat release rate based on user-specified material properties of the combustibles. All three classes of fire models address fully developed fires. The choice of model depends on the reliability of fuel and thermal property data and the desired degree of accuracy.

For compartments with size, aspect ratio, and contents that are comparable to those used for full-scale experiments, the empirical models are appropriate tools.

Zone models are appropriate for a well-stirred combustion chamber. Some zone models, such as COMPF2 (Babrauskas 1979), were designed specifically to model fully developed fires. Others are more general and used primarily to model smoke and heat movement through structures. These models have been applied with mixed success to the prediction of fully developed conditions (Luo, He, Beck 1997 and Buchanan 2001).

Field models can reproduce flashover conditions (NIST 2005f), but the degree of accuracy depends on the quality of the input data. Validation studies (e.g., US NRC 2007) have been performed to quantify the accuracy of the various types of models. The designer should review the documentation for the individual fire model for references to validation work to judge if the model is appropriate for the task at hand.

An international survey of available computer programs for zone and field model that predict fire and smoke growth and spread can be found in Olenick and Carpenter (2003).

6.2.4.4 Selecting a Predictive Method for Fire Modeling

The steps involved in selecting the appropriate fire exposure curve follow from the selection of the design fire. The designer establishes realistic fire loads and ventilation factors. In some cases, effects of compartment size and wall lining materials can be evaluated and introduced into the selection process. These considerations often lead directly to the selection of the appropriate fire exposure curve.

The challenge for engineers is to ensure that the fire exposure curves are reasonably conservative (i.e., do not underestimate temperatures or durations) and that the design fire scenarios are consistent with the performance objectives for the building. While much attention has been focused on computer modeling in recent years, fire exposure curves and parametric equations are still representative of the state of the art for estimating fire exposure curves for fully developed enclosure fires. SFPE (2004) provides guidance on selecting methods that will provide bounding predictions.

6.3 HEAT TRANSFER ANALYSIS

Heat transfer analyses provide estimates of structural component temperatures as a function of time and location in a compartment.

6.3.1 Heat Transfer Mechanisms

Heat is transferred to the structure from the compartment gases and soot through radiation and convection. The rate of heat transfer by each mechanism can change during the course of a fire.

The surface temperature of an object exposed to a fire can be (1) assumed equal to the gas temperature, or (2) estimated from radiative and convective heat transfer analyses. With the first approach, the temperature history of the object is determined by calculating the conductive heat transfer through its layers of material (e.g., the insulation and steel section of an enclosed column). This approach is relatively simple to apply. However, this simple estimation is unlikely to provide bases to evaluate transient stress states since the rates at which components of composite elements heat also depend on the section surface areas and masses, factors not considered in the approach.

The second approach generally yields higher fidelity. Following is a brief introduction to radiative and convective heat transfer analysis. SFPE (2002, pp. 1-1 to 1-89) and any number of engineering text books (Incropera and DeWitt 2002) provide more thorough explanations.

Thermal radiation often dominates heat transfer from a fire to the solid surfaces of a compartment. The radiative heat transfer is proportional to the difference in the effective temperatures of the fire and the surface to the 4th power. If the fire and the surface are assumed to be perfect (or black body) absorbers and emitters of radiation (an approximation that is surprisingly reasonable), the steady-state rate of heat transferred from the fire to the surface by radiation, Q_r (W), can be approximated by:

$$Q_r = F_{2\text{-}1} \; \sigma \; A_2 \; (T_1^{\;4} - T_2^{\;4})$$

where:

$F_{2\text{-}1}$ Radiative view factor between the surface and the fire
σ Stefan-Boltzmann constant ($5.669 \cdot 10^{-8}$ W/m^2 K^4)
A_2 area of the surface receiving heat from the fire (m^2)
T_1 effective temperature of the fire (K)
T_2 average temperature of the surface (K)

The absorptivity, α, of a real surface is the fraction of the heat absorbed when compared to a perfect surface. In fires, the absorptivity is about equal to its emissivity, ε, and varies with the amount of soot in the flame. Since it is common during fires for soot to increase in concentration and to deposit on the exposed surfaces of the structure, thereby increasing the emissivity of both the fire and the surfaces, the rate of heat transferred by thermal radiation changes.

The radiation view factor accounts for the geometry of the fire and of the surface' It is a measure of the fraction of energy leaving the surface that is exposed to the fire. For example, objects' surfaces directly facing a fire may "see" the entire fire volume and be heated intensely by radiation. However, the back side of exposed objects is shielded and may not receive any heat directly from the fire. Even so, the shielded sides of exposed objects may be heated by radiation from other surfaces, by the hot smoke layer some distance from the fire source, or by convective heat transfer.

Radiation heat transfer can be complex, and the simplified calculations suggested above, while generally more accurate than estimating objects' temperatures solely from gas temperatures, still might not yield high accuracy for complicated exposures.

Convective heat transfer is always present in a structure fire, but it increases in importance relative to radiative heat transfer as the temperature of the environment decreases. The simplest approach to estimating the heat transfer from a gas to a surface due to convection, Q_c, is with the following expression:

$$Q_c = h \, A_s \, (T_s - T_f)$$

where:

h	convective heat transfer coefficient (W/m^2·°C)
A_s	area of the surface exposed to the gas (m^2)
T_s	temperature of the surface (°C)
T_f	temperature of the gas (°C)

The convective heat transfer coefficient is a complex function of the gas velocity, the gas properties, and the gas and surface temperatures. The value for h is generally found to lie between 10 W/m^2-°C and 30 W/m^2-°C (SFPE 2004).

The convective and radiative components are normally additive, and provide the boundary condition for the conduction analysis. The initial temperature, thermal properties, and geometry of the structural components must be specified to determine the temperature, T, everywhere in the structure using the following general equation:

$$\frac{\partial^2 T}{\partial x^2} + \frac{\partial^2 T}{\partial y^2} + \frac{\partial^2 T}{\partial z^2} = \frac{\rho c}{k} \frac{\partial T}{\partial t}$$

where:

T	temperature in the structure (°C)
x, y, z	defines the position in the structure (m)
t	time (s)
k	structural material thermal conductivity (W/m-°C)
c	specific heat of the structural material (J/kg-°C)
ρ	density of the structural material (kg/m^3)

If the heat transfer into the structure is primarily one dimensional and the thermal conductivity is constant, then, under steady-state conditions, the convective and radiative components of heat transfer are balanced by the conduction normal to the surface, and the above equation reduces to

$$-k \frac{dT}{dx} = \frac{Q_c + Q_r}{A}$$

where A is the area of the surface.

The heat transferred to the structure will continue as long as the temperature of the gas is higher than the temperature of the surface. Once the fire burns itself out, heat transfer by surface radiation and convection will cool the structure.

6.3.2 Heat Transfer Analysis Methods

Heat transfer analyses can range from simple one-dimensional (or lumped mass) equations such as those suggested in section 6.3.2 to analyses with finite element software, depending on the complexity of the geometry and heat flow. While some of the most rudimentary analyses (principally, one-dimensional analyses) can be performed by hand, practical and efficient analyses normally require computer-based solutions. Two- and three-dimensional heat transfer analyses are sufficiently complex problems, particularly when heat flows through different materials with dissimilar thermal properties, that they require computer-based solutions. Finite element software can represent the salient characteristics of materials of construction and insulation, including the effects of air gaps and the various modes of heat transfer (radiation, convection, and conduction) in complicated geometries.

Finite element analyses use solid elements for heat transfer analysis. Temperature dependent thermal, strength, and stiffness properties of steel, concrete, and fire protection materials are given in the following sources. Chapter 4 gives concrete properties. Chapter 5 and Harmathy (1983) give steel and fire protection material properties. Temperature dependent properties of structural steels can also be found in NIST 2005b. NIST 2005d has temperature dependent properties of the passive fire protection used in the WTC towers, and NIST 2005e has temperature dependent data for concrete.

For compartment fires, an emissivity of 1.0 and a convective heat transfer coefficient of 30 $W/m^2 \cdot ^\circ C$ can be conservatively assumed, though the value may range between 10 $W/m^2 \cdot ^\circ C$ and 30 $W/m^2 \cdot ^\circ C$ (SFPE 2004).

6.3.3 Temperature Data for Structural Analysis

A heat transfer analysis will produce temperature profiles that may vary across the sections and along the lengths of components. This information usually needs to be converted to input for structural analyses software. Since, the capabilities of various software differ and designers might decide to analyze to levels of precision that allow approximations, the component temperature data need to be reduced to be compatible with the structural response analysis program and the goals of the designer. For instance, the designer might need to approximate non-linear temperature profiles as linear if the analysis program can not receive as input non-linear profiles between nodes along the length or across the member section.

The temperature histories from the heat transfer analysis should be reviewed before selecting a time interval for the structural analysis input of temperature data. For example, a 120 min fire scenario may produce relatively rapid heating in the first 30 min to 60 min, followed by cooling at a slower rate. The time interval for the analysis of the heating portion of the time-temperature relationship should capture the temperature rise through linear interpolation between data points (NIST 2005e). The same time interval could be used for the more gradual temperature changes typically associated with cooling, or the time interval could be increased if the time-temperature curve can be adequately simulated with large time increments.

6.4 ANALYTICAL APPROACHES FOR FIRE-RESISTANT DESIGN

6.4.1 Prescriptive Building Code Requirements

Building codes specify fire-resistance rating requirements (in terms of hours) for performance of building components under exposure to the standard fire. Fire ratings are usually determined by testing (e.g., following ASTM E 119). Catalogues of components with specified construction details and their approved fire rating are available for selecting an appropriate protection configuration that is consistent with the architecture and goals of the project. When this approach is used, designers need to consider connections between rated structural components. For instance, if columns have a required 3 hour rating, and floor beams and girders have a required 2 hour rating, the design may specify that the connections be protected according to the higher rating (i.e., match the column level of protection).

The standard fire approach gives relative levels of protection required for overall building performance. By verifying structural component performance in accordance with standard fires, manufacturers of fire protection products are certifying their products according to a common standard. Performance of components and the structural system during actual fires requires a comparison between actual conditions and conditions during standard fire tests. Codification of performance levels based on exposures to standard fires does not necessarily ensure that the structural system will, indeed, be able to sustain the effects of a real fire for the duration implied by the rating.

6.4.2 Standardized Analytical Approaches

For designs where pre-qualified details based on standard fire testing are not available, analytical approaches for determining fire resistance of many common structural systems are available in reference documents such as *Standard Calculation Methods for Structural Fire Protection* (ASCE/SEI/SFPE 29-99 1999) and *Standard Method for Determining Fire Resistance of Concrete and Masonry Construction Assemblies* (ACI 216.1-97/TMS 0216.1-97 1997). These standard methods provide procedures for determining fire resistance to standard ASTM E 119 fires, in terms of hours, before reaching defined endpoint criteria for common structural components and type of fire protection.

Both ASCE/SEI/SFPE 29-99 and ACI 216.1-97/TMS 0216.1-97 present design information and methods for computing a fire resistance rating for geometries for which there is no rating established by test, and determining an appropriate thickness of insulating materials. These references guide designers on the selection of sprayed fire protection, concrete cover thickness, enclosures of gypsum products, enclosures with masonry or concrete, and other means to add fire resistance to steel components. These references also provide guidance for detailing connections and gaps between components to minimize the potential for fire spread.

Both ASCE/SEI/SFPE 29-99 and ACI 216.1-97/TMS 0216.1-97 provide means to quantify the minimum thickness of concrete for a component to meet required fire resistance ratings. For instance, methods are given for determining the effective thickness of tapered webs in concrete components and the effective thickness of hollow core panels. These methods and others establish a consistent approach for determining a fire resistance rating of untested configurations that are similar to configurations with fire resistance ratings established by ASTM E119 tests.

ASCE/SEI/SFPE 29-99 also provides guidance on fire resistance calculations for steel, wood, and masonry. For steel sections, it specifies how to calculate geometric parameters such as "heated perimeter" and "section factor" of rolled steel sections as a function of flange width, depth, and web thickness for a number of configurations. The section factor is expressed in several ways, but it is essentially the ratio of the heated surface area to the mass (volume). The Manual of Steel Construction for Load and Resistance Factor Design (AISC 2001) has tables that list the section factor ratio as the weight per foot divided by the section perimeter for 4 cases: 3 sides of a steel beam (one flange protected from heat exposure), 4 sides of a steel beam, 3 sides of a box perimeter for a steel beam in an enclosure (one flange protected from heat exposure), and 4 sides of a box perimeter for a steel beam in an enclosure.

Prescribed definitions of geometric parameters allow users to access charts, graphs, and tables in which fire resistance parameters are presented. For instance, ACI 216.1-97/TMS 0216.1-97 lists minimum equivalent thicknesses of concrete walls to achieve specific fire resistance ratings as a function of concrete aggregate type. These ASCE and ACI references also prescribe minimum cover over reinforcing steel and prestressing steel in concrete beams, and other factors that determine fire resistance, to achieve specific resistance ratings relative to the standard fire resistance requirements specified in building codes.

The methods discussed above have the limitation that only the performance of individual components is considered. Structural system performance at elevated temperatures is not addressed.

6.4.3 Computational Analytical Approaches

To determine the structural system response or the structural behavior of components and subsystems to design fires, one or more of the following analyses can be conducted to provide the required data:

- Fire modeling with hand calculations, parametric equations, or computer programs to develop fire exposure curves (e.g., gas temperature vs. time curves) for design fire scenarios.
- Heat transfer analyses to compute temperature histories of structural components.
- Structural analysis with service gravity loads and elevated temperature histories to determine deformations and stresses due to thermal expansion, reduced stiffness and strength, and plastic and creep strains.

Computer programs for fire modeling will give gas temperature-time histories, which are needed as input data for heat transfer analyses to determine temperature profiles through component cross sections. With some software packages, component stresses and deformations can be evaluated simultaneously with the determination of temperatures; with others, temperature profiles are computed separately and then provided as input for structural analyses. Chapters 4 and 5 listed a number of computer programs that can analyze concrete and steel structures.

Typically, if a finite element analysis package offers an option to solve heat transfer and structural response in a single analysis, it uses solid elements. While some software have features that partially automate the analyses for certain mechanical systems, such as engines or radiators, such conveniences generally do not exist for beam and shell elements normally needed for structural analyses. The analyst must first conduct a heat transfer analysis with solid elements and then translate equivalent temperatures to the nodes of beam and shell elements for the structural analysis.

6.4.4 Selecting an Approach

Standard fire, time equivalence, and fire exposure curves and parametric equations are used for designing fire resistance for most typical structures. A performance based approach should be used when building stakeholders (owners, insurers, regulators, etc.) wish to understand how the building will actually perform in fire or wish to minimize risks through structural fire resistant design. If a performance based design approach uses analytical procedures instead of standard fire testing or approved analytical approaches, the design team should involve the building code officials early in the process to determine what the building officials will require for approval of the design.

Recommendation 28 of the NIST investigation of the World Trade Center collapse suggested that appropriate design professionals (fire protection engineers and structural engineers) provide the standard of care when designing structures to resist fires in buildings that employ "innovative structural and fire safety systems" (NIST 2005a). This suggests that methods that consider the actual response of the structure to fire (i.e., performance based approaches) be used in buildings with innovative structural systems.

6.5 FIRE RESISTANT DESIGN CONSIDERATIONS

Fire safety is provided in a building by a combination of active fire suppression and passive fire protection. Active fire suppression includes firefighting and automatic devices, such as sprinklers, to control the spread of fire. Passive fire protection includes measures, such as fire barriers, that control the spread of fire or insulations, such as concrete cover and spray-applied fire protection that delay the effects of fire on the structure. These guidelines address the fire resistance provided by passive fire protection measures that prevent structural collapse in fire. The following approaches can enhance fire resistance in buildings:

- Control fuel quantity and locations.
- Control fire spread.
- Control ventilation characteristics.
- Protect construction materials.

6.5.1 Fuel Control

Designers usually have limited control over the fuels that are built into or brought into buildings. In large measure, the occupancy determines the fuel load inside buildings since much of the fuel in buildings is derived from its contents.

On the other hand, designers, working with their clients, can make certain fuel-related decisions that can impact fire severity. Owners and their consultants should consider carefully the fuel load associated floor, wall, and ceiling finishes to minimize the avoidable potential for fires to ignite, spread, and grow.

For instance, ceramic floor tile and concrete-based floor finishes add no fuel to a building, and they have the added advantage of providing thermal inertia that absorbs heat and reduces fire severity. Painted plasterboard walls over metal studs contribute less fuel to a fire than do wood finishes and many manufactured cubicle partition systems. Similarly, plasterboard ceilings constitute lower fuel load (and better fire barriers) than do certain ceiling tiles.

Owners and their consultants can often control decisions about locations of fuels for power sources. To the extent that such fuels are needed for heating, air conditioning, backup power, or other purposes, options for placement and isolation within facilities and associated risks should be reviewed.

6.5.2 Control of Fire Spread

Effective compartmentation of interior spaces can limit fire spread by creating barriers among spaces. Features of effective compartmentation include fire-rated partitions and proper protection of penetrations through these partitions. Fire-rated partitions prevent the passage of flames, hot gases, and heat transmission to the unexposed side of the barrier. This is typically accomplished by having the barrier extend from floor slab to floor slab, particularly if non-fire rated ceiling tiles are used in the facility.

Additionally, architectural and functional features can effectively impact the potential for fires to spread. For instance, window size, orientation, and spacing from floor to floor can contribute to the fire intensity by influencing how easily fire spreads from one floor to another. Tall windows, closely spaced from floor to floor, in general are more likely to contribute to vertical spread than are smaller windows with deep spandrel panels in between.

6.5.3 Ventilation Control

Because ventilation affects the temperature and duration of compartment fires, selection of ventilation characteristics can be used as part of a design strategy. Most predictive methods use the "ventilation factor" (see 6.2.1.3), the area of the ventilation opening multiplied by the square root of its height, as input. Judicious selection of opening geometry can be used to reduce fire severity in a compartment.

6.5.4 Fire Resistance and Protection of Construction Materials

Most materials of construction require insulation to achieve resistances that are commensurate with performance requirements in building codes. The amount of protection that designers must provide depends on the inherent resistance of the construction materials, geometry of structural components, function in the structural system, and performance objectives for the building.

Fire protection for structural components is accomplished by some form of insulation, usually as applied coatings, encasement of components, and enclosures around components that separates the structural component from the fire environment.

Some protection products and systems maintain a barrier to the transmission of heat by their static structure and form (e.g., enclosing steel columns in a fire-rated enclosure). Other protection means dissipate heat energy by physical or chemical transformations. Physical transformations include release of entrapped moisture (e.g., heat of hydration for bound water in spray applied insulation). Chemical transformations include endothermic decomposition and heat-induced expansion to create insulation layers (e.g., intumescent coatings). Examples of these products and system types are discussed in Section 5.1.2.

The selection of materials for critical structural components will have a direct impact on strategies for providing fire resistance. For example, reinforced concrete components with appropriate detailing can sustain the effects of fire temperatures for relatively long periods if they have an adequate concrete cover over the steel reinforcement (see Sections 4.3.1 and 4.4.3). The concrete cover over the steel reinforcement acts as insulation, and delays temperature rise in the steel reinforcement if it does not spall or crack during the fire event. The dimensions and the thermal conductivity of concrete components may be designed so that substantial time is required for the temperature of the steel reinforcement to rise to damaging levels.

Brick and concrete masonry components are similar in behavior to concrete components: they are relatively large in dimension and tend to insulate any embedded reinforcing steel.

Steel structural components subject to fire exposure generally need protective insulating layers such as spray applied insulation, enclosures, or concrete coatings for fire protection. This is particularly true for lightweight steel systems such as cold-formed steel components and fabricated components such as steel joists, which have large surface-area-to-volume ratios.

190

Some building codes place criteria (i.e., allowable building height and floor area) on buildings as a function of occupancy, combustibility of materials of construction, and levels of protection. In those circumstances, the occupancy and size of a building can affect the suitability of certain materials of construction.

6.6 FIRE-RESISTANT DESIGN OF CONCRETE STRUCTURES

This section discusses design practices for common types of concrete construction and the response of reinforced concrete structural components under exposure to fire, with an emphasis on concrete floor systems. Refer to Buchanan (2001) for worked examples of concrete component designs for fire exposure.

6.6.1 Concrete Floor Beams

A reinforced concrete beam supporting a concrete floor will have positive moment steel positioned near the bottom of the beam at midspan. The top of the beam usually is integral with the concrete slab and is, therefore, substantially isolated from exposure to high temperatures from a fire below. Since conventional flexural design of beams neglects the concrete below the neutral axis, loss in strength in this portion of the concrete component does not have a significant impact on load-carrying capacity. The integrity of the steel reinforcement in this region, on the other hand, is critical. The concrete cover over the steel reinforcement acts as an insulator. Thus, when a concrete beam is exposed to fire, the temperature increases rapidly at the section surfaces but more slowly at the interior. The temperature profile across the section drops substantially within a short distance from the exposed surface, resulting in a highly nonlinear temperature profile. Since concrete has a relatively large heat capacity and density, there can be a significant time lag before the temperature of the steel reinforcement increases. Hence, the temperature of the steel reinforcement lags the concrete surface temperature.

During initial phases of a serious fire, the concrete cover normally will remain intact, providing insulation to the steel reinforcement. However, high temperatures weaken concrete, produce high stresses due to thermal expansion, and generate vapor pressures within the concrete which may cause spalling (see 4.4.5).

If the concrete cover over the reinforcing steel spalls during the progression of a fire, the steel is exposed directly to the fire environment. Without insulation, the steel heats rapidly, with a corresponding reduction in strength and stiffness. Even if the concrete does not spall, the temperature of the steel reinforcement may rise if there is sufficient fire duration (ACI 216.1 1997).

6.6.2 Untested Concrete Component Capacity in Design Fires

When a specific design is not addressed by standard fire test results, and when informed judgment will not allow designers to adapt test results, designers need to conduct analyses to validate fire resistance designs. These analyses can follow two paths:

1. Rely on published charts that show the theoretical temperature at depth into concrete components as a function of duration in standard exposure fires (Section 4.3.1).
2. Conduct heat transfer analyses to determine internal temperature increases as a function of time.

Standard fire exposure charts can be adapted to show approximate theoretical temperatures inside concrete components during realistic fires. Some rules for determining direct relations are available (Section 4.3.2), and designers can rely on fire conversion formulas (Section 2.5.4) that give equivalent standard fire exposure times for realistic fires when fire load, ventilation, and compartment surface materials are known. Heat transfer analyses are useful when standard exposure curves cannot be used and when conversions from realistic fire exposures to standard fire exposures are impractical or insufficiently accurate.

With either approach, the goal is to determine the temperature of the steel reinforcement due to the fire exposure, and to find the conditions and durations at which the load carrying capacity of the concrete component has been reduced so that it no longer supports the design service load. The design service load for analysis of fire effects is usually taken as the "point-in-time" load rather than the full load when such loads control for non-fire conditions, which may be taken as 1.2 DL + 0.5 LL (see Table 2.1).

6.6.3 Restraint of Continuous Concrete Floors

Continuous concrete floors may not fail when the positive moment flexural steel at midspan has inadequate strength for the design service loads. Plastic hinges need to form at the support points in addition to the midspan before a failure mechanism is created. In concrete structures with continuous beams, the steel reinforcement for negative moment resistance is embedded in the slab near the top of the beam, and is likely to be relatively protected from the heat in the fire compartment. Hence, formation of negative moment hinges due to reduction in steel reinforcement capacity usually trails the formation of positive moment hinges.

6.6.4 Thermal Expansion Effects

Fully developed fires in a compartment generate intense heat in the compartment. However, adjacent compartments are usually not significantly heated. As the fire spreads in a building, new areas sequentially become hot as the fire intensifies locally while an area of previously intense fire begins to cool.

In a fire environment, forces generated by restraining thermal expansion may have additional effects on component and system survivability. Differences in temperatures of structural components within a structural system may create restraining forces that affect the load-carrying capacity in structural components. For instance, when a fire first begins to heat structural components, those components will expand and push against adjacent structure. If there is continuity within the structural system and adequate strength in the surrounding structural components to resist forces due to thermal expansion, the heated components may develop significant compressive forces.

In reinforced concrete floor beams, thermally induced compression can initially counteract the effects of reduced strength in steel reinforcement due to heating. As the steel tensile strength is being reduced, thermal expansion effects are generating compressive forces and reducing the demand on the steel reinforcement, prolonging the load-carrying life of the beam. However, thermally induced compressive forces along the beam length may also generate P-Δ moments at the center of the beam, which would increase demand on the steel reinforcement.

6.6.5 Reinforced Concrete Floor Sagging

As steel reinforcement temperatures increase, steel stiffness and strength will reduce and floor beams will begin to sag. This deformation could signal failure to contain the fire in the compartment if the floor sagging is sufficient to cause a breach in the fire barrier (e.g., a gap or opening between the floor and the walls). However, large deformations do not necessarily mean that beams have failed to perform adequately in a fire, particularly if the performance goal is prevention of collapse.

In structural systems with adequate continuity, large deformations and a significant loss of stiffness can generate tensile forces in the floor beams. Once deflections become large, the weakened reinforcing steel usually has adequate strength to support the applied loads with the beam acting primarily as a tension member. For this mechanism to develop, splices in the reinforcing steel need to function, and there must be sufficient continuity and strength in the beam detailing and in the structural system around the bay with large deformations to support the tensile reactions generated at the ends of the beam.

6.6.6 Cooling Phase Effects

Assuming that heated beams do not collapse during the fully developed phase of a fire, they may develop significant additional forces during the cooling phase. Surviving components that previously expanded and sagged under the influence of heat will begin to cool and contract. This contraction pulls inward on the adjacent structure, particularly if components were deformed during the fire event. Floor beams with thermally induced compressive loads that were sufficient to cause plastic strains to develop may experience tension as contraction during cooling reduces, and possible reverses, the compressive forces on adjacent components.

During a fire event, the sequence of reinforcing steel yielding, concrete weakening and spalling, midspan and end supports undergoing large rotations, floor beams developing compressive and tensile forces may cause stresses that exceed the level normally expected in concrete components, particularly for steel reinforcement details at connections.

6.6.7 Design of Reinforced Concrete Components

Conventional design of reinforced concrete components usually does not consider the load-carrying capacity after flexural failure. For unrestrained, determinate components, failure usually is assumed when the first plastic hinge forms. For continuous beams, approaches for determining the benefits gained by compression while restraining component expansion (when connections and the surrounding structure can sustain the forces associated with restraint of expansion) and by redistribution of moments that occurs as hinges form are presented in Chapter 4.

When analyses show that the strength of reinforced concrete cannot provide adequate resistance to the effects of fire, within the limitations of other design constraints, the designers have two principal options: change the configuration of the components or apply methods to modify the design-basis fire.

Configuration changes can take the following forms:

- Add restraint.
- Add continuity.
- Increase concrete cover over reinforcing steel.
- Increase the area of reinforcing steel.

Each of these options adds resistance at a cost. The designer needs to anticipate cost, together with other goals of the design, when selecting the best method or combination of methods.

Normally, the addition of restraint and continuity to systems is costly and potentially disruptive to the intended behavior of structural systems and may affect space usage. Enhancement of concrete cover can usually be accomplished at a modest cost, with the benefit of additional insulation provided for the steel reinforcement. Introduction of additional reinforcing steel (beyond that required for design loads under ambient conditions) to provide reserve moment capacity also can be accomplished at a modest cost. Steel reinforcement can often be increased in diameter or added without changing the profile of reinforced concrete components.

When changes in the design of components does not emerge as the preferred approach to adding fire resistance, the performance of reinforced concrete components can be enhanced with many of the same fire protection methods that are available for structural components of other construction materials: concrete components can be insulated from the harsh fire temperatures. To evaluate these methods, the designer will need to refer to qualifying fire tests or pursue heat transfer and structural response analyses to evaluate the effectiveness of various protection methods.

6.7 FIRE-RESISTANT DESIGN OF STEEL STRUCTURES

This section discusses design practices for common types of steel construction and the response of steel structural components and composite floor systems under exposure to fire. Refer to Buchanan (2001) for worked examples of steel component designs for fire exposure.

6.7.1 Steel Behavior at Elevated Temperatures

As discussed in Chapter 5, unprotected steel components are sensitive to the effects of fire. The relatively high thermal conductivity of steel (at least 27.3 W/m^2K as compared to approximately 1.3 W/m^2K for normal weight concrete) and generally thin proportions of steel components make unprotected steel structures susceptible to rapid heating when exposed to fires. For this reason, steel components in structures usually require passive fire protection.

When structural steel component temperatures exceed 400 °C, the yield strength and modulus of elasticity begin to decrease (Milke 2002, NIST 2005b). As temperatures increase in a steel section, its yield strength and elastic modulus decrease. When the yield strength is reduced to the applied stress level, the steel section will begin to yield (i.e., deform under plastic strains and, possibly, creep strains). At the same time, reduced stiffness will increase deflections.

6.7.2 Composite Floor Behavior at Elevated Temperatures

Heating of a composite floor initially causes thermal expansion, which in turn, subjects the floor section and adjacent framing to compressive loads. If a steel beam acts compositely with a concrete slab, the most highly stressed element of the steel beam is typically the bottom flange. If the bottom flange yields as steel temperatures increase, the neutral axis shifts upward. Since thin webs make only minor contributions to moment resistance, yielding of the bottom flange can result in rapid and significant loss of moment capacity as the cross section yielding progresses rapidly upward through the web.

If a composite section is heated sufficiently, large deformations and sagging may occur. The temperature gradients in the steel beam and slab will induce thermal bowing, where the floor section will bow downward to relieve the differential thermal expansion. The connections and surrounding framing will be subject to compressive loads from thermal expansion.

Consider a section of the composite floor that includes a beam and tributary area of the slab, with a protected beam (i.e., insulated with passive fire protection). If the floor section is heated from fires below, typically the bottom flange will have the highest temperatures, as it heats first, with a temperature gradient through the web to the top flange. The top flange will be considerably cooler (often by several hundred degrees), due to contact with the thermal mass of the slab, which acts as a heat sink. If the bottom flange temperatures reduce the yield strength to equal the applied loads, the composite section will develop plastic strains at the point of highest load, which is often near the midspan of simply supported floors. If the plastic strains result in a plastic hinge forming at the highly loaded section, the floor may 'hang' between supports. If the sagging increases to the point that the composite floor is supporting its loads through tensile loads at its connections, then the floor section is often described as being supported through catenary action. The top flange and steel reinforcement of the slab will attempt to carry the tensile stresses, if the remaining capacity is greater than the supported loads.

Six fire tests of an 8-story steel framed structure with composite floors was carried out by a research team at the Building Research Establishment (BRE) in Cardington, Bedfordshire to: (1) gain understanding of the natural fire resistance of such structures, (2) correlate data and observations with predictive numerical models, and (3) establish a more rational design methodology for steel framed building response to fire (British Steel 1999). There are a number of reports and papers written that summarize the test data, observations, and numerical analysis of the tests. British Steel (1999) and O'Connor (2003) are given as summary papers.

6.7.3 Design of Composite Floor Section

The design of steel components for fire exposure is usually based on the assumption that positive and negative moments are limited to the plastic moment, which is based on a reduced yield strength and stiffness based on the maximum temperature estimates.

The flexural capacity of a composite floor section is determined by calculation of the plastic moment across the steel beam and slab section for the estimated maximum temperature that the steel beam components will reach due to the design fires. When determining the flexural capacity for a thermal condition, one should also consider the possible reduction in the stiffness and strength of the concrete and whether the concrete can support the compressive load (meaning that a hinge has not formed and the beam still has some flexural capacity) or if the concrete will crush (signaling the limit of flexural capacity).

Steel beams connections can be designed to carry increased tensile loads for the condition where hinges develop in a composite floor section and cause tensile loads at the supports. Large deformations associated with loss of stiffness and flexural yielding can allow the full cross section of steel beams to act as tensile elements as long as the connections can sustain the substantial deformations associated with large hinge rotations and the surrounding structure can support the forces induced by tension in the beam.

It is not common to design for loss of flexural capacity. However, conventional design does allow full redistribution of moment between the positive and negative moment regions to account for the formation of full hinges at these locations. For steel components to develop full plastic hinges they must be adequately braced at hinge locations to undergo the associated rotations. In addition, the designer must consider the integrity of the bracing element; it must also be designed with consideration for the effects of temperature on its strength and stiffness.

6.7.4 Passive Fire Protection for Steel Components

Common methods to protect steel components from the effects of fire include spray-on fire resistive materials (SFRM), intumescent coatings, and enclosures of gypsum board, mineral fiberboard, concrete, masonry, or similar materials (Section 5.1.2, Milke 2002). Each of these protection systems provides its insulating function through one or more mechanisms: low thermal conductivity, high heat capacity, heat-absorbing reactions, or formation of insulation layers through expansion.

Each protection system has its set of advantages regarding cost, aesthetics, weight, and ease of installation. The systems with perhaps the longest history of use generally are relatively easy to install by trade personnel. Systems, such as enclosures with gypsum or mineral fiberboard products, are easy to install and relatively lightweight. Non-combustible enclosures of masonry materials are also relatively easy to install on columns (but not beams), but they have a weight premium that should be considered in design. SFRM coatings are lightweight, but generally require an enclosure to conceal their unfinished appearance unless they are applied in areas where finishes need not have high aesthetic qualities. SFRM products are somewhat vulnerable to damage over time, particularly in areas where work may be done by other trades. An inspection and maintenance program by the owner can address any loss in integrity of the SFRM coating.

Some of the less-traditional approaches to fire protection of steel, such as intumescent coatings, are still costly when compared to other approaches and often need to be applied by specialty contractors. Some intumescent coatings can be applied as the final finish for exposed steel components, thereby adding to the steel framing appearance for architectural purposes.

6.8 REFERENCES

ACI 216.1 (1997), *Standard Method for Determining Fire Resistance of Concrete and Masonry Construction Assemblies*, Farmington Hills, Mich.: American Concrete Institute.

Aktan, A.E., Ellingwood, B.R., and Kehoe, B. (2007), *Forum: Performance-Based Engineering of Constructed Systems*, Journal of Structural Engineering, Vol 133, Issue 3, pp. 311-322.

ASCE/SFPE 29, (1999), *Standard Calculation Methods for Structural Fire Protection*, Reston, Va.: American Society of Civil Engineers.

ASTM E119-02 (2002), *Standard Methods of Fire Tests of Building Construction and Materials* West Conshohocken, Pa.: American Society for Testing and Materials.

Babrauskas, V. (1996), *"Fire Modeling Tools for Fire Safety Engineering: Are They Good Enough?"* Journal of Fire Protection Engineering, 8:2 (1996) 87-96.

British Steel Plc. (1999), *The Behaviour of Multi-storey Steel Framed Buildings in Fire*, Swinden Technology Centre, South Yorkshire, United Kingdom.

Buchanan, A.H. (2001), *Structural Design for Fire Safety*, U.K.: John Wiley & Sons, Ltd.

Combustion Science & Engineering, Inc. (2002), *International Survey of Computer Models for Fire and Smoke*, http://www.firemodelsurvey.com

Drysdale, D. (1999), *An Introduction to Fire Dynamics*, 2nd ed., Chichester, U.K.: John Wiley & Sons, Ltd.

Harmathy, T.Z. (1983), *Properties of Building Materials at Elevated Tempertures*, paper NRCC-20956 (DBR-P-1080), Division of Building Research, National Research Council, Canada (1080), pp 72, March 1983.

Huggett, C. (1980), *Estimation of Rate of Heat Release by Means of Oxygen Consumption Measurements*, Fire and Materials, **4,** 61-65

ICC (2006), *ICC Performance Code for Buildings and Facilities*, Country Club Hills, Ill.: International Code Council.

Incropera, F.P. and DeWitt, D.P. (2002), *Fundamentals of Heat and Mass Transfer*, John Wiley and Sons, 2002.

Karlsson, B., and J. Quintiere (1999), *Enclosure Fire Dynamics*, Boca Raton, Fla.: CRC Press.

Law, M. (1983), *A Basis for the Design of Fire Protection of Building Structures*, The Structural Engineer 61A:5 (1983).

Law, M. (1997), *A Review of Formulae for T-Equivalent,* Fire Safety Science Proceedings of the Fifth International Symposium, London: International Association for Fire Safety Science.

Lie, T. (2002), *Fire Temperature–Time Relationships,* SFPE Handbook of Fire Protection Engineering, 3rd ed., Quincy, Mass.: National Fire Protection Association.

Luo, M., He, Y. and Beck, V., *Application of Field Model and Two-Zone Model to Flashover Fires in a Full-Scale Multi-Room Single Level Building,* Fire Safety Journal 29, 1 (1997).

Magnusson, S.E., and S. Thelandersson (1970), *Temperature-time curves of complete process of fire development,* ACTA Polytechnica Scandinavica, Civil Engineering and Building Construction Series No.65, Stockholm.

McGrattan, K., et al. (2002), *Fire Dynamics Simulator (Version 3)—User's Guide*, NISTIR 6784, Gaithersburg, Md.: National Institute of Standards and Technology.

Milke, J. (2002), *Analytical Methods for Determining Fire Resistance of Steel Members*, SFPE Handbook of Fire Protection Engineering, 3rd ed., Quincy, Mass.: National Fire Protection Association.

NIST (2005a), *Federal Building and Fire Safety Investigation of the World Trade Center Disaster: Final Report on the Collapse of the World Trade Center Towers*, NIST NCSTAR 1, Gaithersburg, Md.: National Institute of Standards and Technology.

NIST (2005b), *Federal Building and Fire Safety Investigation of the World Trade Center Disaster: Mechanical Properties of Structural Steels*, NIST NCSTAR 1-3D, Gaithersburg, Md.: National Institute of Standards and Technology.

NIST (2005c), *Federal Building and Fire Safety Investigation of the World Trade Center Disaster: Fire Structure Interface and Thermal Response of World Trade Center Towers*, NIST NCSTAR 1-5G, Gaithersburg, Md.: National Institute of Standards and Technology.

NIST (2005d), *Federal Building and Fire Safety Investigation of the World Trade Center Disaster: Passive Fire Protection*, NIST NCSTAR 1-6A, Gaithersburg, Md.: National Institute of Standards and Technology.

NIST (2005e), *Federal Building and Fire Safety Investigation of the World Trade Center Disaster: Component, Connection, and Subsystem Structural Analysis*, NIST NCSTAR 1-6C, Gaithersburg, Md.: National Institute of Standards and Technology.

NIST (2005f), *Federal Building and Fire Safety Investigation of the World Trade Center Disaster: Computer Simulations of the Fires in the World Trade Center Towers*, NIST NCSTAR 1-5F, Gaithersburg, Md.: National Institute of Standards and Technology.

NFPA (2006), *Building Construction and Safety Code, 2006 Edition,* NFPA 5000, Quincy, Mass.: National Fire Protection Association.

O'Connor, M.A., B.R. Kirby, and D.M. Martin (2003), *Behaviour of a multi-storey composite steel framed building in fire*, Structural Engineer, Vol. 81, no. 2, pp. 27-36, 21 Jan 2003.

Olenick, S.M. and Carpenter, D.J. (2003), *An Updated International Survey of Computer Models for Fire and Smoke*, Journal of Fire Protection Engineering, Vol. 13, No. 2, 87-110 (2003)

Society of Fire Protection Engineers (2002), *Handbook of Fire Protection Engineering, 3rd Edition*, Bethesda, Md.: Society of Fire Protection Engineers.

Society of Fire Protection Engineers (2004), *Engineering Guide: Fire Exposures to Structural Elements*, Bethesda, Md.: Society of Fire Protection Engineers.

Society of Fire Protection Engineers (2006), *Engineering Guide: Fire Risk Assessment*, Bethesda, Md.: Society of Fire Protection Engineers.

Society of Fire Protection Engineers (2007), *SFPE Engineering Guide to Performance-Based Fire Protection*, Bethesda, Md.: Society of Fire Protection Engineers.

US NRC (2007), *Verification and Validation of Selected Fire Models for Nuclear Power Plant Applications*, NUREG 1824, March 2007, Washington, DC

Zicherman, J.B. and Eliahu, A. (1998), *Finish Ratings of Gypsum Wallboards,*, Fire Technology, Vol. 34, no. 4, pp. 356-362, 1998.

www.ingramcontent.com/pod-product-compliance
Lightning Source LLC
Chambersburg PA
CBHW082030190526
45166CB00017B/2270